Studies in Logic

Logic and Argumentation

Volume 93

Truth and Knowledge

Studies in Logic Series Editor
Dov Gabbay   dov.gabbay@kcl.ac.uk

Truth and Knowledge

Karl Schlechta

ISBN 978-1-84890-403-3

College Publications
Scientific Director: Dov Gabbay
Managing Director: Jane Spurr

http://www.collegepublications.co.uk

Contents

Chapter 1

Introduction

1.1 Overview

1.1.1 Background

We adhere to "normal" scepticism, i.e. we are certainly aware of various fallacies of perception etc., but assume that reality exists, that we are conscious, and so is the reader, etc.

Modern science, like physics, medical sciences, etc. are good examples of deep and efficient knowledge about (aspects of) the world.

From these assumptions and "gold standards" of systematic knowledge, we try to investigate other sets of knowledge.

(1) First, we are aware that different areas may have different fallacies and interferences in the observations. E.g., the placebo/nocebo effects are important in medicine, but not in physics.

(2) In some areas, it seems impossible to have direct access to phenomena, I have no access to your consciousness, and vice versa, still, we can communicate about our experiences. It is unclear how such questions can be approached in problems about consciousness of animals. We would have to extrapolate. (The author thinks that we should not take an easy way out, in the style of "everything is a bit conscious ....".)

There is nothing mysterious about such situations. If the ideal (gold standard, "absolute truth") cannot be or is too difficult to achieve, we have to be pragmatic and not give up. If we have no freeway, we have to take back country roads.

(3) Physics works with well structured knowledge, we may add, multiply, compare reals etc. Sometimes, our knowledge is less complete and structured, we may have only a partial order, but still would like to do probability theory, so we need suitable approximations. This is e.g. the case in legal reasoning.

(4) If we have no more than perhaps contradictory data, we can still try to come to a reasonable conclusion, based on majorities and past reliability. If a source of data was often wrong in the past, we should be more sceptical than for data from a more reliable source.

(5) It is human that researchers tend to try to confirm their own theories, so we have to confirm/disprove results independently - as far as possible. (In questions about consciousness, this is often impossible.)

(6) Finally, our knowledge should be consistent, and have a possible solution.

In (partial) summary:

- different areas have different traps and fallacies (e.g. placebo effects), to identify them is part of the game.

- if we cannot be as good as in physics, this no reason to give up, we have to do the best we can do, without direct access to data (consciousness problems), partial knowledge, etc.

We also have to be clear about our aims. A successful philosophical analysis of a notion need not mean that we actually think as described in the analysis.

The present text addresses some of above issues, and others.

Remark 1.1.1

To the author's knowlege, there is no systematic overview of the issues, techniques, and solutions, in epistemology and philosophy of science for different areas, as hinted at above.

It might be a book that should be written.

1.1.2   Details

This text looks at problems of truth and knowledge from different angles. The subject of truth and knowledge binds the chapters together, otherwise, they are mostly independent from each other. Their choice is due to the author's interests, and his limited competence.

- We might have insufficient knowledge for certain operations (e.g. comparisons) - how can we approximate sufficient knowledge by a best guess? (Chapter 2 (page 7)).

- Are changes from one case to another relevant for a certain question? (Chapter 5 (page 77)).

- Can we discern honest arguments from propaganda? (Chapter 6 (page 105)).

- To which extent corresponds a formal philosophical analysis to actual thinking about a problem? (This is a meta-question, applicable in many situations.) (Chapter 3 (page 45)).

- And, on a more specific level, we use well known approaches, e.g. to the analysis of counterfactuals, to an analysis of analogical reasoning (Chapter 4 (page 69)), and, finally, we look at Yablo's paradox, analyse his construction, and generalize it to arbitrary formulas of the type $\bigvee \bigwedge \phi_{i,j}$. The latter is an attempt to to come closer to a characterisation of Yablo-like constructions (Chapter 7 (page 117)).

Thus, in other words, the main subjects of this text are:

(1) Generalization of concepts and operations, like distance and size, to situations where they are not definable in the usual way.

(2) A pragmatic theory of handling information (and contradictions) using reliability of the information sources.

(3) Relation of formal semantics to brain processes.

(4) Remarks on Yablo's coding of the liar paradox in infinite acyclic graphs.

From another perspective, we treat

(1) aspects of human reasoning and their formal analoga, and their discrepancies (Chapter 4 (page 69)) and (Chapter 3 (page 45)),

(2) "softening" of formal constructions to better fit some requirements of various situations (Chapter 2 (page 7)),

(3) meta-properties of formal reasoning (Chapter 5 (page 77)),

(4) assessment of information using reliability of its sources (Chapter 6 (page 105)), and

(5) comments on the Yablo paradox (Chapter 7 (page 117)).

In more detail:

(1) Chapter 2 (page 7) generalizes usual operations to structures with weaker properties. In Section 2.2 (page 9) and in Section 2.3 (page 19) we generalize set operations to subsets of the powerset which are not closed under those operations. In Section 2.4 (page 26) we use the height of a element in a partial order to determine the size of that element, and apply our ideas in Section 2.5 (page 31) to the problems seen in Chapter 6 (page 105).

Section 2.8 (page 38) gives, among other things, our motivation to discuss the generalizations of the present chapter.

(2) Chapter 3 (page 45) discusses the (very probable) difference between the simple beauty of the Stalnaker/Lewis semantics for counterfactuals and what happens in our brain, when working with counterfactuals.

Section 3.5 (page 60) gives a short and very simplified picture of the structure of, and processes in the brain.

(3) Chapter 4 (page 69) presents a highly abstract approach to analogical reasoning, in the spirit of (generalized) distance. This is in the spirit of e.g. the philosophical analysis of counterfactual conditionals, as done by Stalnaker and Lewis.

(4) The main contribution of Chapter 5 (page 77) is a detailed examination of the size relation between sets based on filters and ideals - and thus on nonmonotonic logics - of different strengths. Such size relations are used in Chapter 2 (page 7). (The author has discussed other aspects of this problem in other books.)

(5) Chapter 6 (page 105) presents a theory of truth, describing how we can solve conflicts between contradictory information by assigning a dynamic reliability to information sources. We need here generalized operations discussed in Section 2.5 (page 31) of Chapter 2 (page 7).

(6) Finally, Chapter 7 (page 117) presents formal comments and ideas about a solution of the representation problem for Yablo-like structures.

In particular, Chapter 2 (page 7) and Chapter 6 (page 105) may be seen as examples of how to try to find truth in less than perfect situations. In other such situations, we may need different approaches and techniques. We should see ourselves as detectives who will use all clues at hand to find truth.

The formal material in Chapter 2 (page 7) through Chapter 6 (page 105) is largely elementary.

1.2   Acknowledgements

The author would like to thank Andre Fuhrmann, Dov Gabbay, and David Makinson for many very valuable discussions.

Chapter 2

Operations on Partial Orders

2.1 Introduction

2.1.1 Motivation

In reasoning about complicated situations, e.g. in legal reasoning, see for instance [Haa14], the chapter on legal probabilism, classical probability theory is often criticised for imposing comparisons which seem arbitrary. Our approach tries to counter such criticism by a more flexible approach.

We do not have "the best solution", we rather present some suggestions, first, how to work within one partial order, then, how to associate to an element in a partial order in a reasonable way a (rational) number, often in the interval $[0,1]$, so comparisons over different partial orders are possible, as well as operations between partial orders, like multiplication, etc.

The ideas, as well as the formal results, are elementary, and only meant as suggestions.

2.1.2 Overview

(1) Boolean operators:

We discuss in Section 2.2 (page 9) and Section 2.3 (page 19) possibilities to approximate the result of the usual operations of sup, inf, etc. in partial orders which are not complete under these operations.

(1.1) In Section 2.2 (page 9), we first give the basic definitions, see Definition 2.2.3 (page 10), they are quite standard, but due to incompleteness, the results may be sets of several elements, and not single elements (or singletons). This forces us to consider operators on sets of elements, which sometimes complicates the picture, see Definition 2.2.4 (page 12). We then discuss basic properties of our definitions in Fact 2.2.6 (page 14).

(1.2) An alternative definition for sets is given in Definition 2.2.5 (page 18), but Fact 2.2.7 (page 18) shows why we will not use this definition.

(1.3) In Section 2.3 (page 19), we discuss in preliminary outline a (new, to our knowledge) approach, by adding supplementary information to the results of the operations, which may help further processing. The operators now do not only work on elements or sets of elements, but also the additional information, e.g., instead of considering $X \sqcap Y$, we consider $inf(X) \sqcap inf(Y)$, $sup(X) \sqcap sup(Y)$, etc., where "inf" and "sup" is the supplementary information.

(2) Height, size, and probability:

Section 2.4 (page 26) discusses ways to associate size with elements in partial orders, so we can compare them, calculate probabilities of such elements, etc. There are different ways to do this, the "right" way probably depends on the context. This section is related to Section 5.2 (page 80) in Chapter 5 (page 77), where we discussed size comparison in a non-monotonical setting. The approach here is more abstract, the relation is supposed to be given.

(2.1) In Section 2.4.1 (page 26), we introduce the "height" of an element, as the maximal length of a chain from $\perp$ to that element, see Definition 2.4.1 (page 26). We also define relative height, a value between 0 and 1.

(2.2) In Section 2.4.2 (page 28), we argue that the situation for sequences of partial orders may be more complicated than their product - basically as a warning about perhaps unexpected problems.

(2.3) Section 2.4.3 (page 29) introduces two notions of size for sets of elements, one, Definition 2.4.3 (page 29), by the maximal height of its elements, the other, in Remark 2.4.6 (page 31), as the sum of their heights. We also discuss some basic properties of these definitions.

(3) In Section 2.5 (page 31), we discuss the operations needed in Chapter 6 (page 105).

(4) Section 2.7 (page 37) presents some other remarks on abstract operations.

(5) Finally, we discuss our initial motivation for the present chapter in criticism by S. Haack on the use of probabilities in legal reasoning. This is done in Section 2.8 (page 38), where we present some general remarks on questions of ethics, too.

2.2 Boolean Operations in Partial Orders

We define here Boolean operations on not necessarily complete partial orders, and then probability measures on such orders.

In a way, this is a continuation of work in [Leh96] and [DR15].

First, a general remark:

Remark 2.2.1

We are not perfectly happy with our generalizations of the usual operations of $\sqcap$, $\sqcup$, and $\ominus$ to not necessarily complete partial orders. We looked at a few alternative definitions, but none is fully satisfactory.

There are a number of possible considerations when working on a new definition, here a generalization of a standard definition:

- Do we have a clear intuition?

- Is there a desired behaviour?

- Are there undesirable properties, like trivialisation in certain cases?

- Can we describe it as an approximation to some ideal? Perhaps with some natural distance?

- How does the new definition behave for the original situation, here complete partial orders, etc.?

2.2.1 Framework

Assume a finite partial order $(\mathcal{X}, <)$ with TOP, $\top$, and BOTTOM, $\bot$, and $\bot < \top$, i.e. $\mathcal{X}$ has at least two elements. $<$ is assumed transitive. We do not assume that the order is complete.

We will not always detail the order, so if we do not explicitly say that $x < y$ or $y < x$, we will assume that they are incomparable - with the exception $\bot < x < \top$ for any x, and transitivity is always assumed to hold.

2.2.2 Basic Definitions

Definition 2.2.1

(1) For $x, y \in \mathcal{X}$, set $x \uparrow y$ iff $a \leq x$ and $a \leq y$ implies $a = \bot$.

(2) For $X \subseteq \mathcal{X}$, define
$$min(X) := \{x \in X : \neg \exists x' \in X.x' < x\}$$
$$max(X) := \{x \in X : \neg \exists x' \in X.x' > x\}$$

(3) For $Y \subseteq \mathcal{X}$, define
$$sup(Y) := min(\{y' : \forall y \in Y.y \leq y'\})$$
$$inf(Y) := max(\{y' : \forall y \in Y.y \geq y'\})$$
If $sup(Y)$ (or $inf(Y)$) is a singleton, we also write $SUP(Y)$ (or $INF(Y)$).

Fact 2.2.2

$x \uparrow y,\ x' \leq x \rightarrow x' \uparrow y.$

(Trivial by transitivity.)

We define

Definition 2.2.2

(1) $X^y := \{x \in X.x \leq y\}$

(2) $X \leq Y$ iff $\forall x \in X \exists y \in Y.x \leq y$

(3) $X < Y$ iff $X \leq Y$ and $\exists y \in Y \forall x \in X^y.x < y$

Remark 2.2.3

(1) $X \leq \{\top\}$ (trivial).

(2) $X \subseteq Y \Rightarrow X \leq Y$ (trivial).

(3) The alternative definition:

$X \leq_1 Y$ iff $\forall y \in Y \exists x \in X.x \leq y$

does not seem right, as the example $X := \{a, \top\}$, $Y := \{b\}$, and $a < b$ shows, as then $X \leq_1 Y$.

We want to define analogues of the usual boolean operators, written here $\sqcap$, $\sqcup$, $\ominus$.

We will see below that the result of a simple operation will not always give a simple result, i.e. an element (or a singleton), but a set with several elements as result. Consequently, we will, in the general case, have to define operations on sets of elements, not only on single elements. Note that we will often not distinguish between singletons and their element, what is meant will be clear from the context.

2.2.3   Definitions of the Operators $\sqcap, \sqcup, \ominus$

Definition 2.2.3

(1) Let $x, y \in \mathcal{X}$. The ususal $x \sqcap y$ might not exist, as the order is not necessarily complete. So, instead of a single "best" element, we might have only a set of "good" elements.

Define

(1.1) $x \sqcap y := \{a \in \mathcal{X} : a \leq x \text{ and } a \leq y\}$

This is not empty, as $\perp \in x \sqcap y$.

If $X \subseteq \mathcal{X}$ is a set, we define

$\sqcap X := \{a \in \mathcal{X} : a \leq x \text{ for all } x \in X\}$.

In particular, $x \sqcap y \sqcap z := \{a \in \mathcal{X} : a \leq x, a \leq y, a \leq z\}$.

(1.2) We may refine, and consider

$x \sqcap' y := max(x \sqcap y)$

Usually, also $x \sqcap' y$ will contain more than one element.

We will consider in the next section a subset $x \sqcap'' y$ of $x \sqcap' y$, but $x \sqcap'' y$ may still contain several elements.

(2) Consider now $\sqcup$. The same remark as for $\sqcap$ applies here, too.

Define

(2.1) $x \sqcup y := \{a \in \mathcal{X} : a \geq x \text{ and } a \geq y\}$. Note that $\top \in x \sqcup y$.

If $X \subseteq \mathcal{X}$ is a set, we define

$\sqcup X := \{a \in \mathcal{X} : a \geq x \text{ for all } x \in X\}$.

In particular, $x \sqcup y \sqcup z := \{a \in \mathcal{X} : a \geq x, a \geq y, a \geq z\}$.

(2.2) Next, we define

$x \sqcup' y := min(x \sqcup y)$

Again, we will also define some $x \sqcup'' y \subseteq x \sqcup' y$ later.

(3) Consider now $\ominus$.

Define

(3.1) Unary $\ominus$

(3.1.1) $\ominus x := \{a \in \mathcal{X} : a \uparrow x\}$, note that $\perp \in \ominus x$.

If $X \subseteq \mathcal{X}$ is a set, we define

$\ominus X := \{a \in \mathcal{X} : a \uparrow x \text{ for all } x \in X\}$

(3.1.2) Define

$\ominus' x := max(\ominus x)$

$\ominus' X := max(\ominus X)$

Again, we will also define some $\ominus'' x \subseteq \ominus' x$ later.

It is not really surprising that the seemingly intuitively correct definition for the set variant of $\ominus$ behaves differently from that for $\sqcap$ and $\sqcup$, negation often does this. We will, however, discuss an alternative definition in Definition 2.2.5 (page 18), (3), and will show in Fact 2.2.7 (page 18), (3), that it seems inadequate.

(3.2) Binary $\ominus$ We may define $x - y$ either by $x \sqcap (\ominus y)$ or directly:

(3.2.1) $x \ominus y := \{a \in \mathcal{X} : a \leq x \text{ and } a \uparrow y\}$, note again that $\perp \in x \ominus y$, and

(3.2.2) $x \ominus' y := max(x \ominus y)$

For a comparison between direct and indirect definition, see Fact 2.2.6 (page 14), (4.4).

We turn to the set operations, so assume $X, Y \subseteq \mathcal{X}$ are sets of elements, and we define $X \sqcap Y$, $X \sqcup Y$.

One idea is to consider all pairs (x, y), $x \in X$, $y \in Y$ so we define (in contrast to above Definition 2.2.3 (page 10)) for $\sqcap$ and $\sqcup$:

Definition 2.2.4

We define the set operators:

(1) $\sqcap$

    (1.1) $\sqcap$
$$X \sqcap Y := \bigcup \{x \sqcap y : x \in X, y \in Y\}$$

    (1.2) $\sqcap'$
$$X \sqcap' Y := max(X \sqcap Y)$$

(2) $\sqcup$

    (2.1) $\sqcup$
$$X \sqcup Y := \bigcup \{x \sqcup y : x \in X, y \in Y\}$$

    (2.2) $\sqcup'$
$$X \sqcup' Y := min(X \sqcup Y)$$

(3) $\ominus$

    $\ominus X$ and $\ominus' X$ were already defined. We do not define $X \ominus Y$, but see it as an abbreviation for $X \sqcap (\ominus Y)$.

See Definition 2.2.5 (page 18) and Fact 2.2.7 (page 18) for an alternative definition for sets, and its discussion.

2.2.4   Properties of the Operators $\sqcap, \sqcup, \ominus$

We now look at a list of properties, for the element and the set versions.

Fact 2.2.4

Consider $\mathcal{X} := \{\bot, a, b, \top\}$ with $a \uparrow b$. We compare $\sqcap$ with $\sqcap'$, $\sqcup$ with $\sqcup'$, and $\ominus$ with $\ominus'$.

(1) $\top \sqcap a = \{x : x \leq a\} = \{\bot, a\}$, so $\top \sqcap a \neq \{a\}$, but "almost", and $\top \sqcap' a = max(\top \sqcap a) = \{a\}$.

(2) $\bot \sqcup a = \{x : x \geq a\} = \{\top, a\}$, so $\bot \sqcup a \neq \{a\}$, but "almost", and $\bot \sqcup' a = min(\bot \sqcup a) = \{a\}$.

(3) $\ominus a = \{\bot, b\}$, $\ominus' a = \{b\}$, and by Definition 2.2.3 (page 10), (3.1.1), $\ominus \ominus a = \{\bot, a\}$, and $\ominus' \ominus' a = \{a\}$.

(4) Consider $X = \{a, b\} \subseteq \mathcal{X}$.

Then by Definition 2.2.4 (page 12), (1), $\top \sqcap X = \{a, \bot\} \cup \{b, \bot\} = \{a, b, \bot\}$, and $\top \sqcap' X = max(\top \sqcap X) = X$.

(5) Consider again $X = \{a, b\} \subseteq \mathcal{X}$.

Then by Definition 2.2.4 (page 12), (2), $\bot \sqcup X = \{a, \top\} \cup \{b, \top\} = \{a, b, \top\}$, and $\bot \sqcup' X = min(\bot \sqcup X) = X$.

Thus, $\sqcap'$, $\sqcup'$, $\ominus'$ seem the better variants.

We first show some simple facts about the $\leq$ relation for elements and sets (as defined in Definition 2.2.2 (page 10)), and the operators $\sqcap$, $\sqcup$, $\ominus$.

Fact 2.2.5

(1) $\subseteq$

$$X \subseteq X' \Rightarrow X \leq X'$$

(2) $\sqcap$

(2.1) $x \leq x' \Rightarrow x \sqcap y \subseteq x' \sqcap y$

(2.2) $x \leq x' \Rightarrow x \sqcap y \leq x' \sqcap y$

(2.3) $X \subseteq X' \Rightarrow X \sqcap Y \leq X' \sqcap Y$

(2.4) $X \leq X' \Rightarrow X \sqcap Y \leq X' \sqcap Y$

(3) $\sqcup$

(3.1) $x \leq x' \Rightarrow x' \sqcup y \leq x \sqcup y$

(3.2) $X \subseteq X' \Rightarrow X \sqcup Y \leq X' \sqcup Y$

(3.3) Neither
$$X \leq X' \Rightarrow X \sqcup Y \leq X' \sqcup Y$$
nor
$$X \leq X' \Rightarrow X' \sqcup Y \leq X \sqcup Y$$
holds

(4) $\ominus$

(4.1) $x \leq x' \Rightarrow \ominus x' \leq \ominus x$

(4.2) $X \subseteq X' \Rightarrow \ominus X' \leq \ominus X$

(4.3) $X \leq X' \Rightarrow \ominus X' \leq \ominus X$

Proof

(1) $\subseteq$

By definition of $\leq$

(2) $\sqcap$

    (2.1) $x \sqcap y = \{a : a \leq x \wedge a \leq y\} \subseteq \{a : a \leq x' \wedge a \leq y\} = x' \sqcap y.$

    (2.2) By (1) and (2.1)

    (2.3) $X \sqcap Y = \bigcup \{x \sqcap y : x \in X, y \in Y\}.$
           $X \subseteq X' \Rightarrow X \sqcap Y \subseteq X' \sqcap Y \Rightarrow X \sqcap Y \leq X' \sqcap Y.$

    (2.4) $X \leq X' \Rightarrow \forall x \in X \exists x' \in X'.x \leq x'.$
           $X \sqcap Y = \bigcup \{x \sqcap y : x \in X, y \in Y\}.$
           Let $x \sqcap y \in X \sqcap Y$, then there is $x' \in X'.x \leq x'$, and $x' \sqcap y \in X' \sqcap Y$,
           but $x \sqcap y \subseteq x' \sqcap y$, so $X \sqcap Y \subseteq X' \sqcap Y$, so $X \sqcap Y \leq X' \sqcap Y,.$

(3) $\sqcup$

    (3.1) By $x \leq x'$ $x' \sqcup y \subseteq \{a : a \geq x \wedge a \geq y\} = x \sqcup y.$

    (3.2) Analogous to (2.3).

    (3.3) Consider $\mathcal{X} = (\bot, a, b, c, \top)$, $b < a$, $X := \{b\}$, $X' := \{a, c\}$, so $X \leq X'$,
           and $Y := \{b, c\}.$
           Then $X \sqcup Y = (b \sqcup b) \cup (b \sqcup c) = \{b, a, \top\}$, and $X' \sqcup Y = (a \sqcup b) \cup (a \sqcup$
           $c) \cup (c \sqcup b) \cup (c \sqcup c) = \{a, \top, \top, \top, c, \top\} = \{a, c, \top\}.$

(4) $\ominus$

    (4.1) $x \leq x'$, so by Fact 2.2.2 (page 10), $a \uparrow x' \Rightarrow a \uparrow x.$
           $\ominus x' = \{a : a \uparrow x'\} \subseteq \{a : a \uparrow x\} = \ominus x.$

    (4.2) $X \subseteq X'$, $a \uparrow x$ for all $x \in X'$, so $a \uparrow x$ for all $x \in X.$
           Thus $\ominus X' = \{a : a \uparrow x$ for all $x \in X'\} \subseteq \{a : a \uparrow x$ for all $x \in X\} =$
           $\ominus X.$

    (4.3) $a \in \ominus X' \Rightarrow a \uparrow x'$ for all $x' \in X'$. Let $x \in X$, then there is $x' \in X'.x \leq$
           x', as $a \uparrow x'$, and $x \leq x'$ $a \uparrow x$. Thus $a \in \ominus X$, so $\ominus X' \subseteq \ominus X.$

We now examine the properties of $\sqcap$, $\sqcup$, and $\ominus$.

Fact 2.2.6

Commutativity of $\sqcap$ and $\sqcup$ is trivial. We check simple cases like $\top \sqcap x$, $\bot \sqcup x$, show
that associativity holds, but distributivity fails. Concerning $\ominus$, we see that $\ominus \ominus x$
is not well-behaved, and neither is the combination of $\ominus$ with $\sqcup$.

(1) $\sqcap$ and $\sqcap'$

(1.1) $\top \sqcap x = x$?

$\top \sqcap x = \{a \in \mathcal{X} : a \leq x\}$.

$\top \sqcap' x = \{x\}$.

(1.2) $\top \sqcap X = X$?

$\top \sqcap X = \{a : a \leq x \text{ for some } x \in X\}$

$\top \sqcap' X = max(X)$ - which is not necessarily X (if there are $a, a' \in X$ with $a < a'$).

(1.3) $x \sqcap x = x$?

$x \sqcap x = \{a \in \mathcal{X} : a \leq x\}$

$x \sqcap' x = \{x\}$.

(1.4) $X \sqcap X = X$?

$X \sqcap X = \bigcup \{x \sqcap y : x, y \in X\}$.

Note that $x \sqcap y \subseteq x \sqcap x$ for all $x, y \in \mathcal{X}$, thus $X \sqcap X = \bigcup \{x \sqcap x : x \in X\}$
$= \{a \in \mathcal{X} : a \leq x \text{ for some } x \in X\}$.

$X \sqcap' X = max(X)$ - which is not necessarily X.

(1.5) $x \sqcap y \sqcap z = x \sqcap (y \sqcap z)$?

Let $A := x \sqcap y \sqcap z = \{a : a \leq x, a \leq y, a \leq z\}$

Set $B := y \sqcap z = \{b : b \leq y, b \leq z\}$

(1.5.1) $\sqcap$

We have to show $A = x \sqcap B$.

$x \sqcap B = \bigcup \{x \sqcap b : b \in B\}$ by Definition 2.2.4 (page 12), (1).

If $a \in A$, then $a \in B$, moreover $a \leq x$, so $a \in x \sqcap a \subseteq x \sqcap B$.

Let $a \in x \sqcap B$, then there is $b \in B$, $a \in x \sqcap b$. As $b \in B$, $b \leq y$, $b \leq z$, so $a \leq x$, $a \leq b \leq y$, $a \leq b \leq z$, so $a \in A$ by transitivity.

(1.5.2) $\sqcap'$

(This just due to the fact that $max(max(A) \cup max(B)) = max(A \cup B)$.)

Set $A' := max(A)$, $B' := max(B)$, so $x \sqcap' B' = max(\bigcup \{x \sqcap' b : b \in B'\})$.

Let $a \in A' \subseteq A \subseteq B$, so there is $b' \geq a$, $b' \in B'$, and by $a \in A$, $a \leq x$, so $a \in x \sqcap b'$.

Suppose there is $a' \in \bigcup \{x \sqcap' b : b \in B'\} \subseteq x \sqcap B$, $a' > a$. Then by (1.5.1) $a' \in A$, contradicting maximality of a.

Conversely, let $a \in max(\bigcup \{x \sqcap' b : b \in B'\}) \subseteq x \sqcap B$, then $a \in A$ by (1.5.1). Suppose there is $a' > a$, $a' \in x \sqcap' y \sqcap' z$, so we may assume $a' \in A'$, then $a' \in max(\bigcup \{x \sqcap' b : b \in B'\})$, as we just saw, contradiction.

Thus, it works for $\sqcap'$, too.

(2) $\sqcup$ and $\sqcup'$

(2.1) $\bot \sqcup x = x$?

$\bot \sqcup x = \{a \in \mathcal{X} : a \geq x\}$.

$\bot \sqcup' x = \{x\}$.

(2.2) $\perp \sqcup X = X$?

$\perp \sqcup X = \{a : a \geq x \text{ for some } x \in X\}$

$\perp \sqcup' X = min(X)$ - which is not necessarily X.

(2.3) $x \sqcup x = x$?

$x \sqcup x = \{a \in \mathcal{X} : a \geq x\}$.

$x \sqcup' x = \{x\}$.

(2.4) $X \sqcup X = X$?

$X \sqcup X = \bigcup\{x \sqcup y : x, y \in X\}$.

Note that $x \sqcup y \subseteq x \sqcup x$ for all $x, y \in \mathcal{X}$, thus $X \sqcup X = \bigcup\{x \sqcup x : x \in X\}$ $= \{a \in \mathcal{X} : a \geq x \text{ for some } x \in X\}$.

$X \sqcup' X = min(X)$ - which is not necessarily X.

(2.5) $x \sqcup y \sqcup z = x \sqcup (y \sqcup z)$?

Let $A := x \sqcup y \sqcup z = \{a : a \geq x, a \geq y, a \geq z\}$

Set $B := y \sqcup z = \{b : b \geq y, b \geq z\}$

(2.5.1) $\sqcup$

We have to show $A = x \sqcup B$.

$x \sqcup B = \bigcup\{x \sqcup b : b \in B\}$ by Definition 2.2.4 (page 12), (2).

If $a \in A$, then $a \in B$, moreover $a \geq x$, so $a \in x \sqcup a \subseteq x \sqcup B$.

Let $a \in x \sqcup B$, then there is $b \in B$, $a \in x \sqcup b$. As $b \in B$, $b \geq y$, $b \geq z$, so $a \geq x$, $a \geq b \geq y$, $a \geq b \geq z$, so $a \in A$ by transitivity.

(2.5.2) $\sqcup'$

(See above comment.)

Set $A' := min(A)$, $B' := min(B)$. $x \sqcup' B' = min(\bigcup\{x \sqcup' b : b \in B'\})$.

Let $a \in A' \subseteq A \subseteq B$, so there is $b' \leq a$, $b' \in B'$, and by $a \in A$, $a \geq x$, so $a \in x \sqcup b'$.

Suppose there is $a' \in \bigcup\{x \sqcup' b : b \in B'\} \subseteq x \sqcup B$, $a' < a$. Then by (2.5.1) $a' \in A$, contradicting minimality of a.

Conversely, let $a \in min(\bigcup\{x \sqcup' b : b \in B'\}) \subseteq x \sqcup B$, then $a \in A$ by (2.5.1). Suppose there is $a' < a$, $a' \in x \sqcup' y \sqcup' z$, so we may assume $a' \in A'$, then $a' \in min(\bigcup\{x \sqcup' b : b \in B'\})$, as we just saw, contradiction.

Thus, it works for $\sqcup'$, too.

(3) Distributivity for $\sqcap, \sqcup, \sqcap', \sqcup'$

Let $\mathcal{X} := \{\perp, x, y, z, \top\}$.

(3.1) $x \sqcap (y \sqcup z) = (x \sqcap y) \sqcup (x \sqcap z)$?

Then $y \sqcup z = \{\top\}$, so $x \sqcap (y \sqcup z) = \{x, \perp\}$.

$y \sqcup' z = \{\top\}$, so $x \sqcap' (y \sqcup' z) = max(\{x, \perp\}) = \{x\}$.

$x \sqcap y = x \sqcap z = \{\perp\}$, so $(x \sqcap y) \sqcup (x \sqcap z) = \mathcal{X}$.

$x \sqcap' y = x \sqcap' z = \{\perp\}$, so $(x \sqcap' y) \sqcup' (x \sqcap' z) = min(\mathcal{X}) = \{\perp\}$.

So distributivity fails for both versions.

(3.2) $x \sqcup (y \sqcap z) = (x \sqcup y) \sqcap (x \sqcup z)$?

$y \sqcap z = \{\bot\} = y \sqcap' z.$

$x \sqcup \bot = \{x, \top\},\ x \sqcup' \bot = \{x\}.$

$x \sqcup y = \{\top\} = x \sqcup' y,\ x \sqcup z = \{\top\} = x \sqcup' z.$

$\top \sqcap \top = \mathcal{X},\ \top \sqcap' \top = \{\top\}.$

So it fails again for both versions.

(4) $\ominus$ and $\ominus'$

(4.1) $\ominus\top = \bot$?

$\ominus\top = \{\bot\}$

$\ominus'\top = \{\bot\}$

(4.2) $\ominus\bot = \top$?

$\ominus\bot = \mathcal{X}$

$\ominus'\bot = \{\top\}$

(4.3) $\ominus\ominus x = x$?

Consider $\mathcal{X} := \{\bot, x', x, y, \top\}$ with $x < x'$.

Then $\ominus x = \{\bot, y\}$, $\ominus'x = \{y\}$, $\ominus(\ominus x) = \{\bot, x', x\}$, $\ominus(\ominus'x) = \{\bot, x', x\}$, $\ominus'(\ominus'x) = \{x'\}$, so it fails for both versions.

(4.4) $x \sqcap (\ominus y) = x \ominus y$?

$x \ominus y = \{a \in \mathcal{X} : a \leq x \text{ and } a \uparrow y\}.$

$\ominus y = \{a \in \mathcal{X} : a \uparrow y\}.$

$x \sqcap (\ominus y) = \bigcup\{x \sqcap a : a \in \mathcal{X}, a \uparrow y\} = \{b \in \mathcal{X} : b \leq x \text{ and } b \leq a \text{ for some } a \in \mathcal{X}, a \uparrow y\} = \{b \in \mathcal{X} : b \leq x \text{ and } b \uparrow y\}$ by Fact 2.2.2 (page 10).

(4.5) $x \sqcap (\ominus x) = \bot$?

$x \sqcap (\ominus x) := \bigcup\{x \sqcap y : y \in (\ominus x)\} = \bigcup\{x \sqcap y : y \uparrow x\}.$ Let $a \in x \sqcap y$ for $y \uparrow x$, then $a \leq x$ and $a \leq y$, so $a = \bot$.

(4.6) $X \sqcap (\ominus X) = \bot$?

$X \sqcap (\ominus X) := \bigcup\{x \sqcap y : x \in X, y \in (\ominus X)\} = \bigcup\{x \sqcap y : x \in X, y \uparrow x' \text{ for all } x' \in X\}.$ Conclude as for (4.5).

(4.7) $x \ominus x = \bot$?

$x \ominus x = \{a \in \mathcal{X} : a \leq x \text{ and } a \uparrow x\} = \{\bot\} = x \ominus' x.$

(4.8) $x \sqcup (\ominus x) = \top$?

Consider $\mathcal{X} := (\bot, a, b, c, ab, \top)$, with $a < ab$, $b < ab$.

Then $\ominus a = \{b, c, \bot\}$, and $a \sqcup (\ominus a) = \bigcup\{a \sqcup b, a \sqcup c, a \sqcup \bot\} = \{ab, \top, a\} \neq \{\top\}.$

$\ominus'a = \{b, c\}$, $a \sqcup' (\ominus'a) = min(\{ab, \top\} \cup \{\top\}) = \{ab\} \neq \{\top\}$, so it fails for both versions.

(4.9) $\ominus$ is antitone: $X \subseteq X' \Rightarrow \ominus X' \subseteq \ominus X$

$a \in \ominus X' \Rightarrow a \uparrow x$ for all $x \in X'$, so $a \uparrow x$ for all $x \in X \Rightarrow a \in \ominus X.$

(4.10) $X \subseteq \ominus \ominus X$

$\ominus X := \{a : a \uparrow x \text{ for all } x \in X\}.$

Let $x \in X$, $a \in \ominus X$. By $a \in \ominus X$, $x \uparrow a$, so $x \in \ominus \ominus X$.

(4.11) $\ominus \ominus X \subseteq X$ fails in general.

Consider $\mathcal{X} := \{\bot, a, b, c, d, \top\}$ with $d < b$, $d < c$.

Then $\ominus \{b\} = \{\bot, a\}$, $\ominus\{\bot, a\} = \{\bot, b, c, d\}$, so $\ominus \ominus \{b\} \not\subseteq \{b\}$.

Definition 2.2.5

We define alternative set operators, and argue in Fact 2.2.7 (page 18) below that they do not seem the right definitions.

(1) $\sqcap_1$, $\sqcap_2$

    (1.1) $X \sqcap_1 Y := \bigcap\{x \sqcap y : x \in X, y \in Y\}$,

    (1.2) $X \sqcap_2 Y := \sqcap(X \cup Y)$

(2) $\sqcup_1$, $\sqcup_2$

    (2.1) $X \sqcup_1 Y := \bigcap\{x \sqcup y : x \in X, y \in Y\}$,

    (2.2) $X \sqcup_2 Y := \sqcup(X \cup Y)$

(3) $\ominus_1$

    $\ominus_1 X := \{a \in \mathcal{X} : a \uparrow x \text{ for some } x \in X\}$

Fact 2.2.7

Consider again $\mathcal{X} := \{\bot, a, b, \top\}$ with $a \uparrow b$, and $X = \{a, b\} \subseteq \mathcal{X}$, and compare to Fact 2.2.4 (page 12):

(1) $\sqcap_1$, $\sqcap_2$ applied to $\top$, X

    (1.1) $\sqcap_1$:

        Then by Definition 2.2.5 (page 18), (1.1), $\top \sqcap_1 X = \{a, \bot\} \cap \{b, \bot\} = \{\bot\}$.

    (1.2) $\sqcap_2$:

        Then by Definition 2.2.5 (page 18), (1.2), $\top \sqcap_2 X = \sqcap\{a, b, \top\} = \{\bot\}$.

(2) $\sqcup_1$, $\sqcup_2$ applied to $\top$, X

    (2.1) $\sqcup_1$:

        Then by Definition 2.2.5 (page 18), (2.1), $\bot \sqcup_1 X = \{a, \top\} \cap \{b, \top\} = \{\top\}$.

    (2.2) $\sqcup_2$:

        Then by Definition 2.2.5 (page 18), (2.2), $\bot \sqcup_2 X = \sqcup\{a, b, \bot\} = \{\top\}$.

(3) $\ominus_1$

(3.1) $\perp \in X \Rightarrow \ominus_1 X = \mathcal{X}$. (Trivial)

(3.2) In particular, $\ominus_1 \mathcal{X} = \mathcal{X}$, which seems doubtful.

(3.3) $X \subseteq \ominus_1 \ominus_1 X$:

Let $X \neq \emptyset$.

By $\perp \in \ominus_1 X$ and the above, $\ominus_1 \ominus_1 X = \mathcal{X}$.

(3.4) $\ominus_1 \ominus_1 X \subseteq X$ fails in general.

Consider $\mathcal{X} := \{\perp, a, b, c, d, \top\}$ with $d < b$, $d < c$.

Then $\ominus_1\{b\} = \{\perp, a\}$, $\ominus_1\{\perp, a\} = \mathcal{X}$, so $\ominus_1 \ominus_1 \{b\} \not\subseteq \{b\}$.

(3.5) $\ominus_1$ so defined is not antitone

Consider $\mathcal{X} := \{\perp, \top\}$, $X := \{\top\}$, $X' := \mathcal{X}$, then $\ominus_1 X = \{\perp\}$, $\ominus_1 X' = \{\perp, \top\}$.

Thus, the variants in Definition 2.2.5 (page 18) do not seem adequate.

2.3 Elements and Sets with a Sign

2.3.1 Basic Idea

We will outline here a - to our knowledge, new - approach, and code the last operation into the result, so the "same" result of two different operations may look differently, and the difference will be felt in further processing the result.

Basically, we give not only the result, as well as we can, but also an indication, what the intended result is, "what is really meant", the ideal - even if we are unable to formulate it, for lack of an suitable element.

More precisely, if the result is a set X, but what we really want is $sup(X)$, which does not exist in the structure $\mathcal{X}$, we will have the result with the "sign" sup, i.e., $sup(X)$, likewise inf and $inf(X)$, and further processing may take this into consideration.

In a way, it is a compromise. The full information gives all arguments and operators, the basic information gives just the result, we give the result with an indication how to read it.

The problem is not due to sets (instead of singletons) as rsults, as the following example shows.

Example 2.3.1

Consider $\mathcal{X} : \{a, b, c, d, e, f, g, x\}$ with $g < c < x < a$, $g < d < x < b$, $g < f < x$, $f < e$.

$a \sqcap b = x = c \sqcup d$, $a \sqcap b \sqcap e = f$, $(c \sqcup d) \sqcap e = f$, $(c \sqcap e) \sqcup (d \sqcap e) = g \sqcup g = g$.

If we note that we always went downward (as in the case $a \sqcap b$ etc.), then the result seems robust, whereas in the second case $(c \sqcup d)$, we go first upward, then downward, and distributivity fails. Thus, adding a sign to x could be used as a warning for suitable further processing.

Consider now Example 2.3.2 (page 20), illustrated by Diagram 2.3.1 (page 23), for $\sqcap'$ and $\sqcup'$. Note that $SUP\{x, x'\}$ and $INF\{x, x'\}$ do not exist, but they are "meant".

$a \sqcap' b$ is not a single element, $\{x, x'\}$ is the best we have, but what we really mean is something like $SUP\{x, x'\}$. Of course, we could memorize the arguments, a and b, and the operation, $\sqcap'$, but then we have no result, and things become complicated when processing. So, we memorize $\{x, x'\}$, but add the "sign" that $SUP\{x, x'\}$ was meant.

Likewise, $c \sqcup' d$ is not a single element, but again $\{x, x'\}$ is the best we have, but this time, we mean rather $INF\{x, x'\}$.

Example 2.3.2

Let $\mathcal{X} := \{a, b, c, d, x, x', y, e, e', f, f'\}$ with

$e < c < x < a < f,$

$e < d < x' < b < f,$

$c < x' < a,$

$d < x < b,$

$e < y < f,$

$e' < x' < f',$

$e' < y < f',$

see Diagram 2.3.1 (page 23). (The relations involving $\bot$ and $\top$ are not shown in the diagram, $\bot \neq e$, $\top \neq f$.)

Consider now the slightly modified Example 2.3.3 (page 20), illustrated by Diagram 2.3.2 (page 24), for $\ominus'$.

Here, $\ominus' y$ is not a single element, but once again $\{x, x'\}$, and what is "really meant" is $SUP\{x, x\}$ - which is absent again.

Example 2.3.3

Let $\mathcal{X} := \{a, b, c, d, x, x', y, e, f\}$ with

$e < c < x < a < f,$

$e < d < x' < b < f,$

$c < x' < a,$

$d < x < b,$

$y < a,$

$y < b,$

see Diagram 2.3.2 (page 24). (The relations involving $\bot$ and $\top$ are not shown in the diagram, $\bot \neq e$, $\top \neq f$.)

Thus:

(1) In Example 2.3.2 (page 20), we have

$a \sqcap' b = \{x, x'\}$, more precisely $a \sqcap' b = sup\{y : y \leq a, y \leq b\} = SUP\{x, x'\}$
- which does not exist, but we do as if, i.e., we give a "label" to $\{x, x'\}$.

Reason:

We have $x < a, b$, $x' < a, b$, $SUP\{x, x'\}$ is the smallest z such that $z > x$, $z > x'$, thus $z < a$, $z < b$, but this z does not exist.

(2) Again in Example 2.3.2 (page 20), we have

$c \sqcup' d = \{x, x'\}$, more precisely $c \sqcup' d = inf\{y : y \geq c, y \geq d\} = INF\{x, x'\}$,
- which does not exist, but we do as if, i.e., we give a "label" to $\{x, x'\}$.

Reason:

We have $x > c, d$, $x' > c, d$, $INF\{x, x'\}$ is the biggest z such that $z < x$, $z < x'$, thus $z > c$, $z > d$, but this z does not exist.

(3) In Example 2.3.3 (page 20), we have

$\ominus' y = sup\{x : x \uparrow y\} = SUP\{x, x'\}$.

To summarize, we have $x, x' \leq sup\{x, x'\} \leq a, b$ and $c, d \leq inf\{x, x'\} \leq x, x'$, and $x \uparrow y$, $x' \uparrow y$, $sup\{x, x'\} \uparrow y$ - but $inf\{x, x'\}$ and $sup\{x, x'\}$ need not exist.

More precisely, we have

Fact 2.3.1

(1) $z \leq inf\{x, x'\} \Leftrightarrow z \leq x$ and $z \leq x'$
(Trivial.)

(2) $z \leq sup\{x, x'\} \Leftrightarrow z \leq x$ or $z \leq x'$
(Trivial.)

(3) $z \geq inf\{x, x'\} \Leftrightarrow z \geq x$ or $z \geq x'$
(Trivial.)

(4) $z \geq sup\{x, x'\} \Leftrightarrow z \geq x$ and $z \geq x'$
(Trivial.)

(5) $z \uparrow x$ or $z \uparrow x' \Rightarrow z \uparrow inf\{x, x'\}$, but not conversely:
$\Rightarrow$: $z \not\uparrow inf\{x, x'\} \Rightarrow z \not\uparrow x$ and $z \not\uparrow x'$ by (1).
Counterexample for the converse: Consider x, x', $inf\{x, x'\}, z, u, u'$ with $inf\{x, x'\} \leq x$, $inf\{x, x'\} \leq x'$, $u < x$, $u < z$, $u' < x'$, $u' < z$. (And $\top, \bot$, of course.) Here, $x \not\uparrow z$, $x' \not\uparrow z$, but $inf\{u, u'\}$ does not exist, so $z \uparrow inf\{x, x'\}$.

(6) $z \uparrow sup\{x, x'\} \Rightarrow z \uparrow x$ and $z \uparrow x'$, but not conversely:
$\Rightarrow$: Trivial.
Counterexample for the converse: Consider $x, x', sup\{x, x'\}, u, z$, with $x \leq sup\{x, x'\}$, $x' \leq sup\{x, x'\}$, $u < z$, $u < sup\{x, x'\}$. So $z \uparrow x$, $z \uparrow x'$, but $z \not\uparrow sup\{x, x'\}$.

Thus, there is not necessarily an equivalence for $\uparrow$ (nor for $\ominus'$), though it may hold in some cases, of course.

Consider now again in Example 2.3.2 (page 20), see Diagram 2.3.1 (page 23), $y \sqcap' \{x, x'\}$, $y \sqcup' \{x, x'\}$, $\uparrow \{x, x'\}$, and $\ominus'\{x, x'\}$ to see the different consequences for further operations. The differences are in $\wedge$ vs. $\vee$ (or $\forall$ vs. $\exists$).

(1) $\sqcap'$

    (1.1) $y \sqcap' sup\{x, x'\} = \{z : z < y \wedge (z < x \vee z < x')\} = \{e', e\}$

    (1.2) $y \sqcap' inf\{x, x'\} = \{z : z < y \wedge (z < x \wedge z < x')\} = \{e\}$

(2) $\sqcup'$

    (2.1) $y \sqcup' sup\{x, x'\} = \{z : z > y \wedge (z > x \wedge z > x')\} = \{f\}$

    (2.2) $y \sqcup' inf\{x, x'\} = \{z : z > y \wedge (z > x \vee z > x')\} = \{f', f\}$

(3) $\ominus'$

    (3.1) $\ominus'sup\{x, x'\} = \{a : a \uparrow x \wedge a \uparrow x'\} = \{\bot\}$

    (3.2) $\ominus'inf\{x, x'\} = \{a : a \uparrow x \vee a \uparrow x'\} = \{\bot, e'\}$

Basically, we remember the last operation resulting in an intermediate result, but even this is not always sufficient as the example in Fact 2.2.6 (page 14), (3.1), failure of distributivity, shows: The intermediate results $y \sqcup' z$, $x \sqcap' y$, $x \sqcap' z$ are singletons, so our idea has no influence.

One could, as said, try to write everything down without intermediate results, but one has to find a compromise between correctness and simplicity.

Diagram Inf/Sup $\sqcap', \sqcup$

Recall that $INF\{x, x'\}$ *and* $SUP\{x, x'\}$ *do not exist*

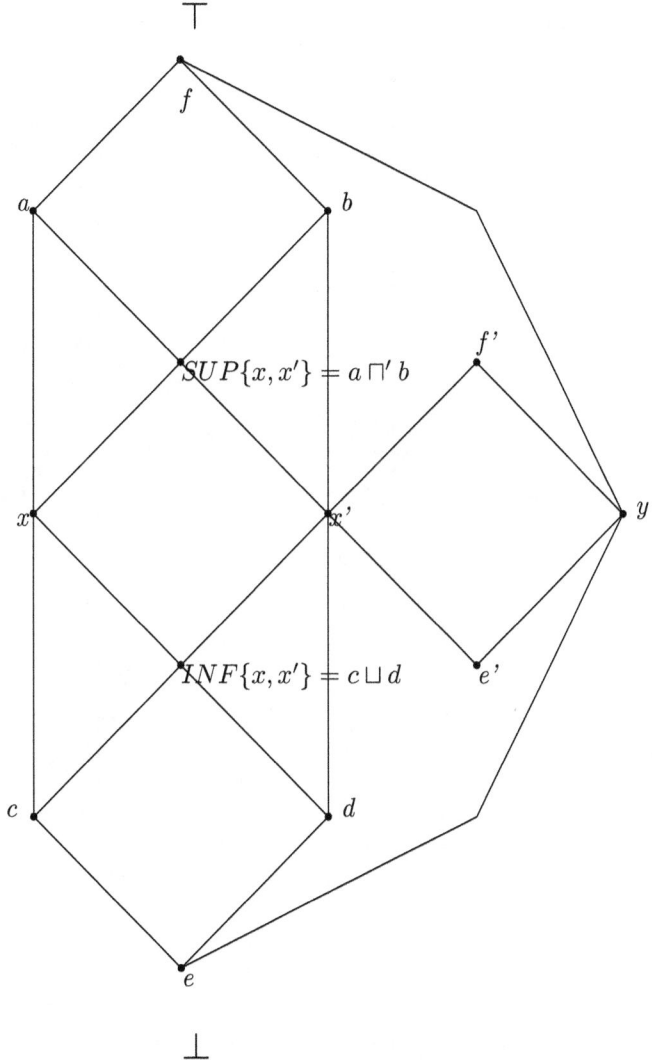

Diagram 2.3.1

Diagram Inf/Sup $\ominus'$

Recall that $INF\{x, x'\}$ and $SUP\{x, x'\}$ do not exist

$\top$

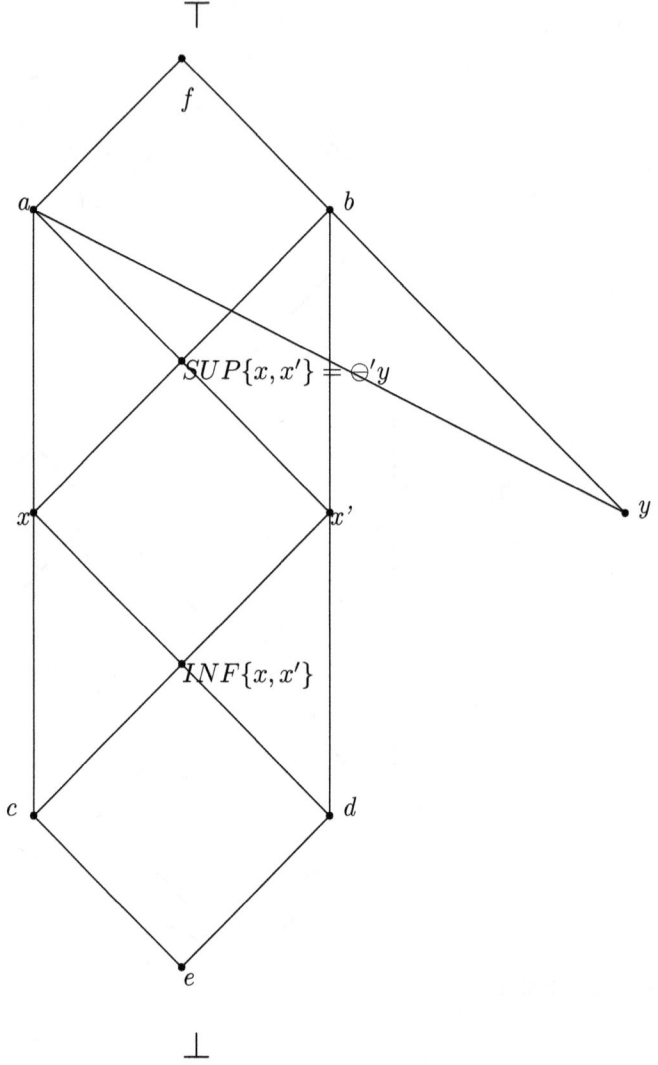

f

a

b

$SUP\{x, x'\} = \ominus'y$

x

x'

y

$INF\{x, x'\}$

c

d

Diagram 2.3.2

e

$\perp$

Remark 2.3.2

The definitions using sup and inf are intuitively better than the old definitions, i.e. those without sign.

However, even if they sometimes give better results than the old definitions, they still do not always conform to the usual result in complete partial orders - and this is probably irredeemably so, the correct results simply are not there.

As an example, we re-consider the example in (4.8) of Fact 2.2.6 (page 14):

$$x \sqcup (\ominus x) = \top?$$

(1) According to the old definition

Consider $\mathcal{X} := (\bot, a, b, c, ab, \top)$, with $a < ab$, $b < ab$.

Then $\ominus a = \{b, c, \bot\}$, and $a \sqcup (\ominus a) = \bigcup \{a \sqcup b,\ a \sqcup c,\ a \sqcup \bot\} = \{ab, \top, a\} \neq \{\top\}$.

$\ominus' a = \{b, c\}$, $a \sqcup' (\ominus' a) = min(\{ab, \top\} \cup \top) = \{ab\} \neq \{\top\}$, so it fails for both versions.

(2) According to the new definition

Consider $\mathcal{X} := (\bot, a, b, c, ab, \top)$, with $a < ab$, $b < ab$.

In the new definition, $\ominus a = sup\{b, c, \bot\}$, and $a \sqcup \ominus a = inf\{y : y \geq a \wedge y \geq b \wedge y \geq c \wedge y \geq \bot\} = \{\top\}$.

The new definition, however, fails for the following example: $\mathcal{X} := \{\bot, y, x, b, \top\}$ with $y < b$, $x < b$. Then $\ominus x = \{y, \bot\}$, $x \sqcup \ominus x = inf\{z : z \geq x \wedge z \geq y \wedge z \geq \bot\} = inf\{b, \top\}$.

2.4   Height and Size in Finite Partial Orders

We assume here a finite, strict, transitive partial order $\mathcal{O}$, with relation $<$. By abuse of language, $\mathcal{O}$ will also be used for the set of elements of $\mathcal{O}$.

Bottom ($\bot$) and top ($\top$) need not exist, neither $\sqcap$ or $\sqcup$, etc. When we write those symbols, we assume that the elements do exist.

As in Definition 2.2.1 (page 9) $\uparrow$ will be used to say that two elements are incomparable: $x \uparrow y$ iff there is no z, $z < x$ and $z < y$, or only $\bot < x$ and $\bot < y$.

2.4.1   Basic Definitions

Definition 2.4.1

(1) Let $x \in \mathcal{X}$.

Set $ht(x) :=$ the length of the longest chain from $\bot$ to x - where we count the number of $<$ in the chain. (If $\bot$ does not exist, take a descending chain, beginning in x, of maximal length.)

(2) This definition might seem arbitrary, why counting from the bottom, and not from the top? And, why counting from bottom or top, not both from bottom and top?

Thus, we introduce an alternative definition for $ht(x)$:

Let $b(x) :=$ the length of the longest chain from $\bot$ to x, $t(x) :=$ the length of the longest chain from x to $\top$, and

$ht(x) := \frac{b(x)}{b(x)+t(x)}$.

Thus, $0 \leq ht(x) \leq 1$, so, when adequate, we may interpret this directly as a probability.

Of course, this might be imprecise, so we may introduce a measure of precision, e.g. for $c(x) :=$ the number of elements in $\mathcal{X}$ comparable to x,

$pr(x) := \frac{c(x)}{card(\mathcal{X})}$.

(See Remark 2.4.3 (page 28), (3) for an alternative, and probably better, idea.)

(3) Let $X \subseteq \mathcal{X}$.

Define

$maxht(X) := \{x \in X : \forall x' \in X.ht(x) \geq ht(x')\}$ and
$minht(X) := \{x \in X : \forall x' \in X.ht(x) \leq ht(x')\}$.

Remark 2.4.1

(1) There is probably not a best choice of definition for all situations, our aim is to indicate ways to proceed with incomplete information.

(2) We pursue only the first definition of ht, as it seems the least complicated one, and we mainly want to illustrate the concept here.

(3) Obviously, the height of an element is related to its "size". Consider the powerset $\mathcal{P}(X)$ over X. A subset X' will be bigger, if it sits higher in the $\subset$ relation. But we may also consider the number (or set) of elements in $\mathcal{P}(X)$ below X'. The bigger this set, the bigger X' is.

This gives a different notion of size of an element in a partial order: $size(x) := \{x' : x' < x\}$ or $size(x) := card(\{x' : x' < x\})$.

We may also consider a mixture of both approaches.

In Remark 2.4.6 (page 31), we give an alternative definition of a probability using height.

Fact 2.4.2

(1) $ht(\bot) = 0$, $ht(\top) > 0$.

(2) $ht(x) \leq ht(\top)$ for all $x \in \mathcal{X}$.

(3) We have $x < y \rightarrow ht(x) < ht(y)$ for all $x, y \in \mathcal{X}$.

(4) If x and y are $<$-incomparable, it does not necessarily follow that $ht(x) = ht(y)$.

(This is trivial, as seen e.g. in the example $\mathcal{X} := \{\bot, a, a', b, \top\}$ with $a < a'$, so $ht(a') = 2$, $ht(b) = 1$, and a', b are incomparable.)

(5) $maxht(X) = maxht(max(X))$, $minht(X) = minht(min(X))$

Definition 2.4.2

(1) $x \sqcap'' y := maxht(x \sqcap y)$,
$\quad X \sqcap'' Y := maxht(X \sqcap Y)$.

(2) $x \sqcup'' y := minht(x \sqcup y)$,
$\quad X \sqcup'' Y := minht(X \sqcup Y)$.

(3) $\ominus'' x := maxht(\ominus x)$,
$\quad \ominus'' X := maxht(\ominus X)$,
$\quad x \ominus'' y := maxht(x \ominus y)$.

We might also have chosen $x \sqcap' y$ instead of $x \sqcap y$, etc., by Fact 2.4.2 (page 27), (5).

Example 2.4.1

Consider $\mathcal{X} := \{\bot, a, b, b', c, \top\}$, with $b < b'$.

Then $\ominus c = \{\bot, a, b, b'\}$, $\ominus' c = \{a, b'\}$, and $\ominus'' c = \{b'\}$, so we lose important information, in particular, if we want to continue with Boolean operations.

For this reason, the versions $\sqcap''$, $\sqcup''$, $\ominus''$ should be used with caution.

Remark 2.4.3

(1) We may, for instance, define similarity between two points x and y in a partial order, by the length of the longest common part of paths from bottom to x and y.

(2) Uncertainty of x may be defined by the number of y incomparable with x.

(3) A probably better idea is as follows:

Let X be the set of elements y incomparable with x. Consider a chain $C \subseteq X$, which has maximal length, say L. This gives an idea how much refinement is possible in X (and thus for the position of x itself). The bigger L, the less precise (or certain) the value for x is.

(If we were to learn more about the partial order, we might be able to compare x with all elements in C, so x might be below, or above, or in the middle of the chain C.)

2.4.2   Sequences

Example 2.4.2

In the second example, we compensate a loss in the second coordinate by a bigger gain in the first. Thus, the situation in the product might be more complex that the combined situations of the elements of the product.

(1) Consider $\mathcal{X} := \{0, 1\}$ and $\mathcal{X}' := \{0', 1'\}$ with the natural orders. In $\mathcal{X}$, $ht(1) = 1$, in $\mathcal{X}'$, $ht(1') = 1$.

Order the sequences in $\mathcal{X} \otimes \mathcal{X}'$ by the value of the sequences, defined as their sum.

So in $\mathcal{X} \otimes \mathcal{X}'$, $\perp = (0, 0') < (0, 1') < (1, 1') = \top$, $(0, 0') < (1, 0') < (1, 1')$, and $ht((1, 1')) = 2$.

(2) Consider now $\mathcal{X} := \{0, 2\}$ and $\mathcal{X}' := \{0', 1'\}$ with the natural orders. In $\mathcal{X}$, $ht(2) = 1$, in $\mathcal{X}'$, $ht(1') = 1$ again.

Order the sequences in $\mathcal{X} \otimes \mathcal{X}'$ again by the value of the sequences, defined as their sum.

So in $\mathcal{X} \otimes \mathcal{X}'$, $\perp = (0, 0') < (0, 1') < (2, 0') < (2, 1') = \top$, and $ht((2, 1')) = 3$.

Of course, we use here additional structure of the components, sum and difference.

In general, we may consider rules like:

$\sigma\sigma' < \tau\tau'$ iff $\sigma'\sigma < \tau'\tau$ and

$\sigma\sigma' < \tau\tau'$ iff $\sigma\sigma' < \tau'\tau$ etc.

We might extend the comparison to sequences of different lengths by suitable padding, e.g. $\langle 0, 1, 0 \rangle$ to $\langle 1, 1 \rangle$ by appending (e.g.) 0 to $\langle 1, 1 \rangle$, resulting in $\langle 1, 1, 0 \rangle$.

2.4.3 Probability Theory on Partial Orders Using Height

We define two notions of size of a set here:

(1) the size of a set is the maximal height of its elements, in Section 2.4.3.1 (page 29), and

(2) the size of a set is the sum of the heights of its elements, in Remark 2.4.6 (page 31).

The first can be seen as a "quick and dirty" approach, the second as a more standard one.

There probably is no unique best solution, it will depend on the situation at hand.

2.4.3.1 Size of a Set as Maximal Height of its Elements

Definition 2.4.3

(1) For $X \subseteq \mathcal{X}$, we set $ht(X) := max\{ht(x) : x \in X\}$.

If we are interested in $sup(X)$, we might define $ht(sup(X)) := ht(X) + 1$, and if we are interested in $inf(X)$, we might define $ht(inf(X)) := min\{ht(x) : x \in X\} - 1$.

(2) We may define a relative height by $rht(x) := \frac{ht(x)}{ht(\top)}$, and we have $0 \leq rht(x) \leq 1$, which may be interpreted as the probability of x.

Thus, we define $P(x) := rht(x)$, and $P(X)$ similarly for $X \subseteq \mathcal{X}$.

Fact 2.4.4

We have the following facts for the height for $\sqcap$ and $\sqcup$:

(1) $ht(X \sqcap X') < ht(X), ht(X')$,

(2) $ht(X), ht(X') < ht(X \sqcup X')$

Proof

This is trivial, as any chain to $(X \sqcap X')$ may be continued to a chain to X and X'. The second property is shown analogously. Alternatively, we may use Fact 2.4.2 (page 27), (3).

$\square$

Remark 2.4.5

When we work with subsets of some powerset, we use, unless defined otherwise, $\subsetneq$ for $<$, $\cap$ for $\sqcap'$, $\cup$ for $\sqcup'$, and $\ominus'$ is set complement.

Example 2.4.3

Consider $ht(X)$ as defined in Definition 2.4.1 (page 26), (1).

These examples show that $ht(X)$, $ht(X')$ may be arbitrarily bigger than $ht(X \sqcap X')$, and $ht(X \sqcup X')$ may be arbitrarily bigger than $ht(X)$ and $ht(X')$.

(1) Let $\mathcal{X} := \{\emptyset, X_1, X_2, X_1', X_2', X, X', X \cap X', X \cup X', X''\}$, with
$X := \{a, a', b\}$, $X' := \{b, c, c'\}$, $X_1 := \{a\}$, $X_2 := \{a, a'\}$, $X_1' := \{c\}$, $X_2' := \{c, c'\}$. Thus, $X \cap X' = \{b\} = X''$, and $ht(X \cap X') = 1$, but $ht(X) = ht(X') = 3$.

(2) Let $\mathcal{X} := \{\emptyset, X_1, X_2, X, X', X \cup X'\}$, with
$X := \{a, a'\}$, $X' := \{b, b'\}$, $X_1 := \{a, b\}$, $X_2 := \{a, a', b\}$. Thus, $ht(X \cup X') = 3$, but $ht(X) = ht(X') = 1$.

$\square$

Example 2.4.4

Some further examples:

Consider $ht(X)$ again as defined in Definition 2.4.1 (page 26), (1), and $P(X) := \frac{ht(X)}{ht(\top)}$.

$P(x) + P(\ominus x)$:

$P(x) + P(\ominus x)$ may be 1, but also < 1 or > 1 :

(1) $= 1$:

Consider $\mathcal{X} := \{\bot, a, b, \top\}$.
Then $\ominus a = \{b, \bot\}$, $\ominus' a = \{b\}$, and $P(a) = P(b) = 1/2$.

(2) < 1 :

Consider $\mathcal{X} := \{\emptyset, \{a\}, \{b, d\}, \{a, b, c\}, \{a, b, c, d\}\}$.
Then $\ominus\{a\} = \{\emptyset, \{b, d\}\}$, $\ominus'\{a\} = \{\{b, d\}\}$. Thus $ht(\{a\}) = ht(\ominus'\{a\}) = 1$, $ht\{a, b, c, d\} = 3$, and $P(\{a\}) = P(\ominus'\{a\}) = 1/3$, and $1/3 + 1/3 < 1$.

(3) > 1 :

Consider $\mathcal{X} := \{\emptyset, \{a\}, \{a, a'\}, \{b\}, \{b.b'\}, \{a, a', b, b'\}\}$.
Then $\ominus'\{\{a, a'\}\} = \{\{b, b'\}\}$, and $ht\{a, a'\} = ht\{b, b'\} = 2$, $ht\{a, a', b, b'\} = 3$, so $P(\{a, a'\}) = P(\ominus'\{a, a'\}) = 2/3$, $2/3 + 2/3 > 1$.

We now consider independence, which we may define as usual:

Definition 2.4.4

A and B are independent iff $P(A \sqcap B) = P(A) * P(B)$.

This, however, might be too restrictive, alternatives come to mind, e.g.

Definition 2.4.5

A and B are independent iff

$P(B) := \alpha$ and

$P(A \sqcap B)/P(A) = \alpha,$

or

$P(A \sqcap B)/P(A) < \alpha$, and $P((\ominus A) \sqcap B)/P(\ominus A) \geq \alpha,$

or

$P(A \sqcap B)/P(A) > \alpha$ and $P((\ominus A) \sqcap B)/P(\ominus A) \leq \alpha,$

The best definition might also be domain dependent.

Remark 2.4.6

We turn to a more standard definition of size, based on point measure, where the size of a point x is again $ht(x)$, but the size of a set is now the sum of the sizes of its points.

Alternatively, we may define for $X \subseteq \mathcal{X}$:

(1) $\mu(X) := \Sigma\{ht(x) : x \in X\},$

(2) $P(X) := \frac{\mu(X)}{\mu(\mathcal{X})}$

But we have similar problems as above with this definition, e.g. with $P(x)$ and $P(\ominus x)$, etc.:

If $x = \bot$ or $x = \top$, then $P(x) + P(\ominus x) = 1$, but if $x \neq \bot$, and $x \neq \top$, then $P(x) + P(\ominus x) < 1$, as $\top$ is missing.

Similarly, as $P(x \sqcap \ominus x) = 0$, we will often have $P(x \sqcup \ominus x) \neq P(x) + P(\ominus x) - P(x \sqcap \ominus x)$.

This, however is not due to incompleteness, as we can easily see by considering complete partial orders. Consider e.g. $\mathcal{X} := \{\bot, a, a', b, b', \top\}$ with $a < a'$, $b < b'$. Then $\ominus a' = \{b, b'\}$, and $ht(a') = ht(b') = 2$, $ht(a) = ht(b) = 1$, $ht(\top) = 3$. $P(a') = 2/9$, $P(\ominus a') = 3/9$, $P(a' \sqcap \ominus a') = 0$, $P(a' \sqcup \ominus a') = 3/9$.

Of course, for this definition of P, considering a disjoint cover of $\mathcal{X}$ will have the desired property.

The reader may also consider the ideas in [DR15] on qualitative probability.

2.5   The Mean Value of Sets

We will use the maen value in Chapter 6 (page 105), see also Section 2.6 (page 33).

(1) Bare sets

Suppose we have just sets, without any additional structure, so we may count the elements, and use the set constants like $\emptyset$, and operations like $\cap$, $\cup$, -, and thus also the symmetrical set difference as a measure of distance $A \Delta B :=$ $(A - B) \cup (B\text{-}A)$, or its cardinality.

We now consider some variants.

(1.1) Use of Δ, a candidate for mean value has Δ-minimal distance from A and B.

Let A, B be disjoint, and consider Z with small Δ-distance from A and B. Let $Z := \emptyset$. $\emptyset \Delta A = A$, $\emptyset \Delta B = B$. Let $Z := A \cup B$. $Z \not\supseteq A = B$, $Z \not\supseteq B = A$, so $\emptyset$ and $A \cup B$ are equivalent, which indicates that Δ is not the right idea.

(1.2) Some more examples.

If $A \cap B \neq \emptyset$, then $(A \cap B)\Delta A = A\text{-B}$, $(A \cap B)\Delta B = B\text{-A}$, $(A \cup B)\Delta A =$ $B\text{-A}$, $(A \cup B)\Delta B = A\text{-B}$, $\emptyset \Delta A = A$, $\emptyset \Delta B = B$, so $A \cap B$ and $A \cup B$ are equivalent, $\emptyset$ is worse.

Take A, B, C pairwise disjoint. then $\emptyset \Delta A = A$ etc., $(A \cup B \cup C)\Delta A =$ $B \cup C$ etc., so considering the distances individually (to A, then to B, etc.), $\emptyset$ is better, but taking the union, they are equivalent.

(1.3) Interior and exterior average.

It seems more interesting to consider the interior and exterior distance, where we minimize $Z - A_i$ for the interior average, and $A_i - Z$ for the exterior average.

In this definition, any $Z \supseteq \bigcup A_i$ is optimal for the exterior average, and any $Z \subseteq \bigcap A_i$ is optimal for the interior average. This seems wrong.

(1.4) Combination of interior and exterior average

We suggest a combination of interior and exterior average, e.g. by a lexicographic order (first interior, then exterior, or vice versa), or by counting the interior twice, the exterior once, etc. The choice will probably depend on the intention.

Some examples (we work here with counting the elements of the set differences, not directly with the sets) for $A := \{a, b\}$ and $B := \{b, c\}$:

(1.4.1) $Z := \emptyset$.
$$card(Z - A) = card(Z - B) = 0,$$
$$card(A - Z) = card(B - Z) = 2.$$

(1.4.2) $Z := \{b\}$.
$$card(Z - A) = card(Z - B) = 0,$$
$$card(A - Z) = card(B - Z) = 1.$$

(1.4.3) $Z := \{a\}$.
$$card(Z - A) = 0, card(Z - B) = 1,$$
$$card(A - Z) = 1, card(B - Z) = 2.$$

(1.4.4) $Z := \{a, b\}$.
$$card(Z - A) = 0, card(Z - B) = 1,$$
$$card(A - Z) = 0, card(B - Z) = 1.$$

(1.4.5) $Z := \{a, c\}$.
$card(Z - A) = card(Z - B) = 1$,
$card(A - Z) = card(B - Z) = 1$.

(1.4.6) $Z := \{a, b, c\}$.
$card(Z - A) = card(Z - B) = 1$,
$card(A - Z) = card(B - Z) = 0$.

(1.4.7) $Z := \{a, b, c, d\}$.
$card(Z - A) = card(Z - B) = 2$,
$card(A - Z) = card(B - Z) = 0$.

Thus, by emphasis on the interior, (1.4.2) is the best (better than (1.4.1) by the exterior), by emphasis on the exterior, (1.4.6) is the best, better than (1.4.7) by the interior.

(1.5) Further refinements are possible, e.g.

(1.5.1) We may prefer those Z with more equal distances to the A_i.

(1.5.2) We may consider the square of the distances, thus penalising big differences.

(1.5.3) We may give some sets bigger weight, counting elements (and thus difference) twice.

(1.6) Counting elements

An alternative approach is to divide the numbers of elements by the number of sets, and chosing accordingly some elements from each set. This does not seem very promising. (Problems with division, e.g. 2/3, are probably not very serious.)

(2) Additional structure

The reader may have noticed that the situation for two sets is similar to the semantic version of symmetric theory revision - except that we have no distance.

Suppose we have a distance, so we can generalize symmetic theory revision to more than two sets as follows:

Consider A_i, $i \in I$. Define the average by $\bigcup\{(A_i \mid (\bigcup\{A_j : j \in I, i \neq j\})) : i \in I\}$ where $A \mid B := \{b \in B : \exists a_b \in A \forall b' \in B \forall a' \in A(d(a_b, b) \leq d(a', b'))\}$.

We might use multisets to give more weight to some sets.

2.6 Generalizing the Operations Used in Chapter 6

2.6.1 Operations Used in Chapter 6

In Chapter 6 (page 105), we use the definitions

(1) r_i: value of A_i

(2) ρ_i: reliability of A_i

(3) ρc_i: reliability of communication channel i

(4) m mean value of the r_i

(5) δ_i: distance between m and r_i

(6) δ: mean value of the δ_i

(7) t: number of r_i used to calculate m

We used the following operations in Chapter 6 (page 105)

(1) Operations Variant 1:

    (1.1) m (mean value) of the r_i

    (1.2) δ_i = distance between m and r_i

    (1.3) δ = mean value of all δ_i

    (1.4) adjusting ρ_i using δ, δ_i, and old ρ_i

(2) Operations Variant 2:

    (2.1) adjust r_i using old ρ_i

(3) Operations Variant 3:

    (3.1) multiply old m by t

(4) Operations Variant 4:

    (4.1) put ρ_i in relation to δ_i (is δ_i small in comparison to ρ_i?)

(5) Operations for communication and own reliability:

    (5.1) (serial) combination of two reliabilities, here ρ_i and ρc_i

    (5.2) conversely, break down a modification of a combination of two reliabilities to a modification of the individual reliabilities (this should be an inverse operation to the first operation here)

    (5.3) for $\in_i$, see Chapter 6 (page 105), nothing new

When we generalize from $\Re$ to more general structures, we will probably first treat linear orders, then full power sets, and then general partial orders. We give some examples.

2.6.2 Generalization From Values in $\Re$ to Sets

Ideas (at least in first approach, let ρ_i, ρc_i will be reals in $[0, 1]$, otherwise some calculations seem difficult - if they are not real, we transform them into a real using height etc.):

The following, however, seems an easy generalization for ρ_i :

Reliabilities (of agents or messages) will be multisets of the form $\{r_i a_i : i \in I\}$. r_i will be a real value between -1 and +1, a_i should be seen as a "dimension".

This allows for easy adjustment, e.g. ageing over time, shifting importance, etc., as we will shortly detail now:

- the real values allow arbitrarily fine adjustments, it is not just $\{-1, 0, 1\}$,

- the dimensions allow to treat various aspects in different ways,

- for instance, we can introduce new agents with a totally "clean slate", 0 in every dimension, or preset some dimensions, but not others,

- the uniform treatment of all dimensions in is not necessary, we can treat different dimensions differently, e.g., conflicts between two agents in dimension a_i need not touch dimension a_j, etc.

Thus, we have arbitrarily many dimensions, with possibly different meaning and treatment, and within each dimension arbitrarily many values. This is not a total order, but within each dimension, it is.

(1) Operations for Variant 1:

    (1.1) mean value m : this seems the main problem.

        Consider the "mean value of sets". The best idea might be to count the occurrence of elements in the sets considered, and chose the set of those elements with the best count (or above a certain threshold).

        Alternatively, given a notion of distance between sets, we might chose those elements, which have the smallest distance from the sets considered.

    (1.2) δ_i we consider (a variant of) symmetrical difference Δ

        (1.2.1) δ_i is a number (Δ as number)

        (1.2.2) δ_i is a set (Δ as set)

    (1.3) δ

        (1.3.1) Case (1.2.1): classical

        (1.3.2) Case (1.2.2): as in (1.1)

    (1.4) Adjusting ρ_i :

        (1.4.1) Case (1.2.1): evident.

        (1.4.2) Case (1.2.2): put Δ in relation to U (with $\mathcal{P}(U)$ considered).

(2) Operations for Variant 2:

    (2.1) We should give weight to elements and perhaps non-elements, too.

(3) Operations for Variant 3:

    (3.1) count multiple times

(4) Operations for Variant 4:

    (4.1) transform both to numbers

(5) Operations for communication etc.:

    (5.1) multiplication and breaking down.

2.7 Appendix

2.7.1 The Core of a Set

The following remarks are only abstractly related to the main part of this chapter. The concept of a core is a derivative concept to the notion of a distance, and the formal approach is based on theory revision, see e.g. [LMS01], or [Sch18b], section 4.3.

We define the core of a set as the subset of those elements which are "sufficiently" far away from elements which are NOT in the set. Thus, even if we move a bit, we still remain in the set.

This has interesting applications. E.g., in legal reasoning, a witness may not be very sure about colour and make of a car, but if he errs in one aspect, this may not be so important, as long as the other aspect is correct. We may also use the idea for a differentiation of truth values, where a theory may be "more true" in the core of its models than in the periphery, etc.

In the following, we have a set U, and a distance d between elements of U. All sets X, Y, etc. will be subsets of U. U will be finite, the intuition is that U is the set of models of a propositional language.

Definition 2.7.1

Let $x \in X \subseteq U$.

(1) $depth(x) := min\{d(x,y) : y \in U - X\}$

(2) $depth(X) := max\{depth(x) : x \in X\}$

Definition 2.7.2

Fix some $m \in \mathbf{N}$, the core will be relative to m. One might write $Core_m$, but this is not important here, where the discussion is conceptual.

Define

$core(X) := \{x \in X : depth(x) \geq depth(X)/m\}$

(We might add some constant like $1/2$ for $m = 2$, so singletons have a non-empty core - but this is not important for the conceptual discussion.)

It does not seem to be easy to describe the core operator with rules e.g. about set union, intersection, etc. It might be easier to work with pairs $(X, depth(X))$, but we did not pursue this.

We may, however, base the notion of core on repeated application of the theory revision operator $*$ (for formulas) or $|$ (for sets) as follows:

Given $X \subseteq U$ (defined by some formula ϕ), and $Y := U - X$ (defined by $\neg\phi$), the outer elements of X (those of depth 1) are $Y \mid X$ ($M(\neg\phi * \phi)$). The elements of depth 2 are $(Y \cup (Y \mid X)) \mid (X - (Y \mid X))$, $M(((\neg\phi) \vee (\neg\phi * \phi)) * (\phi \wedge \neg(\neg\phi * \phi)))$ respectively, etc.

We make this formal.

Fact 2.7.1

   (1) The set version

        Consider X_0, we want to find its core.

        Let $Y_0 := U - X_0$

        Let $Z_0 := Y_0 \mid X_0$

        Let $X_1 := X_0 - Z_0$

        Let $Y_1 := Y_0 \cup Z_0$

        Continue $Z_1 := Y_1 \mid X_1$ etc. until it becomes constant, say $Z_n = X_n$

        Now we go back: $Core(X_0) := X_n \cup \ldots \cup X_{n/2}$

   (2) The formula version

        Consider ϕ_0, we want to find its core.

        Let $\psi_0 := \neg\phi_0$

        Let $\tau_0 := \psi_0 * \phi_0$

        Let $\phi_1 := \phi_0 \wedge \neg\tau_0$

        Let $\psi_1 := \psi_0 \vee \tau_0$

        Continue $\tau_1 := \psi_1 * \phi_1$ etc. until it becomes constant, say $\tau_n = \phi_n$

        Now we go back: $Core(\phi_0) := \phi_n \vee \ldots \vee \phi_{n/2}$

2.8   Motivation: Ethics and philosophy of Law

2.8.1   Introduction

We mention here some aspects of philosophy of law, in particular, where they are related to other subjects of this text, or other work by the author.

2.8.2   Haack's Criticism of Probability In Law

These are comments on Susan Haack's criticism of probabilistic approaches to legal reasoning. Page numbers refer to [Haa14]. See [Sta18c], too.

See [Haa14], *p.* 62, in shorthand:

   (a)

        evidential quality is not necessarily linearly ordered.

   (b)

        $p(\phi) + p(\neg\phi) = 1$, but in the case of no or very weak evidence neither might be warranted.

(c)

By the product rule of probability, the probability of combined evidence is never stronger than the individual probabilities are - but combined evidence might be stronger.

We think that (a) is the deepest objection, (b) and (c) are about details of application, and not about applicability in principle of probability theory to evidence. The criticism of (b) and (c) may also be directed against evidence in natural science, and this is obviously wrong.

- On (a).

  Assigning numbers or places in a total order seems indeed rather arbitrary in many cases.

  There are approaches to generalized probability theory which address such questions, see [Leh96], [DR15].

  In addition, nonmonotonic logic can be seen as qualitative reasoning about size, another way of doing generalized probability theory. See here e.g. [GS16], in particular sections 5.3, 6.4, 6.5, and chapter 11 there, and also [Sch95-3].

  This objection was also the initial reason to develop Chapter 2 (page 7).

  We should also mention here that some comparisons might seem unethical, e.g. comparing the "value" of human life, without comitting to equal value. We do not know how to treat this formally.

- On (b).

  This seems the same problem as in Intuitionistic Logic (which can best be seen as constructive logic, one has a proof, a counterexample, or neither).

  Fermat's conjecture was right or wrong. But for a long time, no one had a proof either way.

  It is the difference between what holds, and what we know to hold. This should not be confused.

- On (c).

  This might be the most difficult point, as it has so many interpretations, e.g.

  (1) do we speak about a hypothesis, how a scenario might have developed, e.g. how the perpetrator might have entered the house,

  (2) do we speak about a chain of observations, the more detailed they are, the more they allow to differentiate between hypotheses (compare to an experimentum crucis in science), and the less likely it happened by chance (compare to the 5σ rule in physics),

  (3) do we speak about uncertain observations, which may support each other to a global probability, though taken individually, the exactness of all elements might be relatively weak

  etc. etc.

Moreover, one should separate the quality of an explanation from its likelihood. A detailed explanation is better than a less detailed one, as an explanation, even though it is less likely. "It happened" is very likely, but worthless as an explanation.

Perhaps one should compare only explanations of similar quality by their likelihood.

Probability theory might in these cases be more complicated to apply, or not be the right level of abstraction, without being wrong in principle, I think.

Abstract approaches are important for two reasons: First, they allow to isolate reasoning from arbitrary influences, second, if they work with intermediate results (as, e.g., probabilities), they simplify reasoning.

2.8.3   Remarks on Various Aspects of Philosophy of Law

(1) A basic principle is equality, justice as fairness.

This, however, is a necessary, but not sufficient condition. Consider a society where each year the first child born this year is sacrificed to the gods. We will hardly consider this system as a decent legal system.

(2) First, following Kant, we have to separate aims and things as they are (Sein und Sollen).

This has led some people to reject a possible world semantics (here, the set of "good" worlds) for obligation (deontic logic). We do not share this, and think that a set of possible worlds (or systems of such sets) has no fixed interpretation. It may describe what we think possible, what we think good, etc. We have to be clear about the meaning, but this is a different thing. We may even have several such systems simultaneously, e.g. for different moral or legal systems, they may be indexed, etc. In addition, we may look at coherence properties between different such systems.

(3) The next major distinction is between "natural" law and positive law. The first has to do what we "feel" to be right ("Rechtsempfinden" in German law), it certainly depends on the cultural context, and probably has its roots in animal behaviour and feeling - e.g., animals seems to be able to have a bad conscience. The cultural context might also be a history of past aberrations, which we try to avoid in future. It might be difficult to describe, and we refer the reader to Section 3.5 (page 60), where we discuss the limits of language in the context of neuroscience. Positive law is what is written in legal texts, or established in the tradition of legal reasoning.

Conflicts between natural and positive law are a major subject of the philosophy of law, see e.g. the Radbruch formula:

"The conflict between justice and the reliability of the law should be solved in favour of the positive law, law enacted by proper authority and power, even in cases where it is injust in terms of content and purpose, except for cases where the discrepancy between the positive law and justice reaches a level so unbearable that the statute has to make way for justice because it has to be considered"erroneous law". It is impossible to draw a sharper line of demarcation between cases of legal injustice and statutes that are applicable despite their erroneous content; however, another line of demarcation can be drawn with rigidity: Where justice is not even strived for, where equality, which is the core of justice, is renounced in the process of legislation, there a statute is not just 'erroneous law', in fact is not of legal nature at all. That is because law, also positive law, cannot be defined otherwise as a rule, that is precisely intended to serve justice."

We see here that philosophy of law works (perhaps has to work) with somewhat imprecise notions, in this case with a "distance" between natural and positive law. Distance and size are fundamental notions of non-monotonic reasoning and logics, so a connection between philosophy of law and non-monotonic reasoning seems evident. Recall here also the origins of formal theory revision in legal thinking, [AGM85].

(3.1) Natural law, ethics, and morality

One should perhaps invest more thought to clarify above imprecise notions of natural laws, "Rechtsempfinden", etc.

Our background will be one of restraint - laws should not try to regulate too much.

Consider the law that anyone who insults the prophet (Allah) should be killed. Not everyone believes in Allah, so this law cannot claim universal validity. We have to find a general property which excludes such laws. Perhaps a look at Rawls helps. To exclude cases like "I feel extremely extremely bad if not everyone (except myself) feels very bad" to valuate a political system, he excludes such artificial constructions. In our example, Allah is for someone who is not Muslim as real as Snow White. So, it is an artificial construction, and laws should abstain from working with artificial constructions "out of thin air".

(The strength of convictions is a bad criterion whether to make a principle a law, as religious convictions show.)

Conversely, any law based on ideas and emotions which all people share, perhaps even some animals, might be considered well-founded.

Moreover, ethics is perhaps too burdened with absolute notions like (absolute) good, evil, God, to be a good guide for the more pragmatic law and its philosophy.

A more general comment: we should not work with incremental distances (if x is within the field of law, and y is close to x, then so is y), but always measure from the point of departure, to avoid excesses and paradoxa.

(3.2) The distinction between law and morality. This does not seem to be the same as the distinction between positive and natural law.

We may think some behaviour to be immoral, often in a sexual or religious context, without feeling it is against natural law - if we are tolerant enough to do so.

It may be good measure of tolerance and liberalism of a society, if people can live with this distinction.

A few remarks on moral systems: It might be necessary to differentiate the areas of moral judgements and their relations to laws, consider e.g.:

(3.2.1) judgements about possessions, my house, my garden, my car ....

(3.2.2) cohesion of the family, like marital fidelity

(3.2.3) sexuality, like exhibitionism

(3.2.4) personal insult, other attacks on social status

(3.2.5) racism

(3.2.6) doubts or attacks of religious beliefs
         etc.

(4) The next distinction is between consequentialism and deontologism.

    The first describes good or bad results of actions, the second good or bad actions - as in the Ten Commandements.

    The first may lead in excesses to "ends justify means", where a surgeon is authorized to slaughter one patient to use his organs to save several other patients.

    The second sees some actions as intrinsically bad, condemning also tyrannicide.

    Again, as in the Radbruch formula, a compromise seems necessary, involving again some (abstract) notions of size and distance.

    Analogical reasoning in comparing cases has to be done carefully, so we do not justify excesses.

(5) Case based reasoning, i.e. using prototypes, will not escape above distinction between the quality of actions and results. Its reasoning is a special case of analogical reasoning, see Chapter 4 (page 69).

(6) Critical rationalism sees (positive) laws as experimental. Laws are made to have a certain effect on society, but we cannot be totally sure about this effect (society changes, the behaviour of judges is not totally predictable), they might have to be revised, improved, etc.

(7) "Dignity of men" (Menschenwuerde)

    This is a fundamental notion of the German constitution. It has no clear definition, its meaning has changed over time. In the sense of German constitution every human being has Menschenwuerde, one cannot lose it, not even, Hitler, Stalin, etc. lost it through their acts.

    It seems absurd to base a constitution on an unclear notion, but this also leaves space for development through interpretation - see also Point 1 (page 66) in Section 3.5 (page 60) for nonverbal, implicit knowledge.

(8) Equilibrium

Some legal systems seem to try to preserve a certain equilibrium, if person A has done damage to person B, then person A has to compensate person B. This idea is present in the Ur-Nammu code, the oldest known law code, ca. 2100 - 2050 BC, Sumer.

(Note that a positive disturbance of the equilibrium is not punished: if I give you 100.- (without any motive), then no one will punish me.)

(9) Further remarks on consequentialism

In general, acting with bad consequences is considered worse than not acting to prevent (the same kind of) bad consequences: pushing a person in a wheelchair over a cliff is worse than not preventing him to roll over the cliff.

In the context of autonomous cars, there are many questions concerning consequences. Sometimes, they seem far-fetched, not only because they will seldom arise, but also because human beings might act in unpredictable ways, without incurring any legal punishment. Examples: shall the driver rather kill an old man than a young child, when killing seems inevitable. (Strangely, questions about safety of the computer and communications systems against hacking, which are probably a bigger problem, seem neglected. Likewise, the "horizon" of actions (consequences within the next second are certainly important, those after one year probably not) seems little considered.

Note also that comparing the "value" of the life of a human being with other values is regularly done, also in civil life. Chosing the patient to be given a life saving transplant, or an extremely costly or limited other treatment, is based on criteria like age, chances of survival, etc. It seems difficult to proceed otherwise (chosing randomly does NOT seem better!). Building a motorway may always be done better and safer, but we have to limit the costs somewhere.

Chapter 3

Composing Logics and Counterfactual Conditionals

3.1 Introduction

(1) We re-consider the Stalnaker/Lewis semantics for counterfactual conditionals. E.g., "in situation σ, if ϕ were the case, then ψ would be the case, too". The idea is to change σ minimally so that ϕ holds, and see if then ψ holds. E.g., the sun is shining, but if it were to rain, we would use an umbrella. A non-minimal change might be to consider a place with very strong winds, or we carry objects in both hands, and cannot hold an umbrella, etc.

This somehow suggests that we may consider all possible situations and choose those diverging minimally from σ such that ϕ holds. But we have no table of all situations in our head, instead we have to construct suitable situations (in the episodic memory), combining σ with past experiences, etc. Such constructions will be influenced by frequencies of experiences, etc., so the result is not an objective look-up, but rather a subjective construction and evaluation. The actual reasoning process is more complicated than the Stalnaker/Lewis semantics suggests.

(2) Basic entities and operations

The basic entities are scenarios or pictures. Scenarios have a complicated structure, and no atoms, see Section 3.5 (page 60). Usually, we cannot use them as they are, but have to cut them up, and combine them with other basic scenarios (or their parts). The combination itself may be a complicated process, resulting in a multi-neuron path between two scenarios.

(3) Controle

45

The search for suitable scenarios is (usually) a complex active process, the combination might lead to an error, and backtracking. Attention, experience, desires etc. may be important guiding forces.

In a dream, we may combine the picture of a elefant with that of wings. Controle rules this out.

There will usually not be a unique possible combination, but not all are useful.

3.2   Composition of Logics

Consider e.g. a preferential logic which describes a situation, and now we want the situation to evolve over time (as in [GR17]).

We may construct one, complicated, logic to cover all aspects, or, decompose the overall picture into two or more logics, each covering some aspect or aspects.

This is not a deep philosophical problem, but a practical one. We can compare the situation to a programming task: make one big main program, or decompose the main program into a small main program and several functions (or subroutines). The advantage of decomposition is that smaller parts are easier to understand, check, and re-use in other contexts.

On the downside, there are several problems to solve:

(1) we have to determine how to cut the problem into smaller parts in a useful way,

(2) we have to determine how the parts communicate with each other,

(3) is there a main logic, and a hierarchy of auxiliary logics, or are the tasks (and thus the logics) more equally organised?

(4) will one logic wait for the answer of another logic before it can continue its own task, as it needs the answer, or can it just start the other logic (synchrone vs asynchrone cooperation)?

Suppose we have common variables, and the logics work in parallel. Anything may happen, the behaviour is basically unpredictable. It is like a program for seat reservation or bank transfers without momentary locking the status. The actual behaviour depends on run-time properties (speed etc.).

(5) if one logic is not satisfied with the result of another logic, what will it do? Can it re-start the second logic with different arguments, will it start a third logic? Will one logic present several answers, from which another logic may choose?

(6) if logics L and L' have the same syntactic operator, say $*$, is the meaning precisely the same in both logics?

These problems are usually not trivial, as anyone who ever did more important programming tasks can testify. We might call them problems of "logic engineering".

The first problem is influenced by the type of logics already existing, or which seem relatively easy to construct.

The second problem depends on the necessary communication. An efficient, but often not very clear solution is to use (in programming) common variables, but one tends to forget which parts influence their values. A usually better solution is to mention the communicating variable explicitly, stating if communication is only in one direction, or in both directions. In short, we have to find an efficient, but also safe, interface between different logics. (The interface may also be dynamic: in one situation, we have to use "common variables", in another situation, modifying just a few "variables" might be sufficient.)

It does not seem that there will be many general regularities for such logics, except trivial ones, e.g., if we have specificity, then, if logic L provides a more precise argument to logic L', then L' will probably give a better answer.

3.2.1 Examples: Temporal Logic and Anankastic Conditionals

D. Gabbay, G. Rozenberg, and co-authors discuss in [GR17] combinations of temporal logic with extension-based logics (here argumentation theory) and counting of certain events. It seems that using temporal logic as a "master logic", and the extension based logic, or counting certain events, as secondary logics are useful.

Further examples are (usual) counterfactual conditionals (see below) and anankastic conditionals, see e.g. [Sab14] for the latter. In the Stalnaker/Lewis semantics for counterfactual conditionals, the present and the hypothetical situation are described in the same logic (and language). But this need not be the case. If not, we need a "higher" logic which puts models from different logics together, and chooses the closest ones according to some criterion. In anankastic conditionals, example: "If you want to go London, you have to take the Eurostar train.", we have an initial situation S, a desired final situation S', and a means to go from S to S', M, describing an action. Here, the initial situation is, implicitly, some place in western Europe. This will be described in some logic L. The destination will be described in some, perhaps different, logic L'. The part describing the means is probably the most complicated one. Of course, one might first fly to Alaska, and then to London, the suggested choice is supposed to be the simplest, cheapest, etc. But even if we are in Paris at the start, the Eurostar might not always be the best solution. If we live close to Charles de Gaulle airport, and need to go close to Heathrow airport, flying might well be more convenient. So, the choice of the means or action is probably the most complicated part of the reasoning.

3.3   Human Reasoning and Counterfactual Conditionals

We started our investigation by looking into the semantics of counterfactual conditionals (CFC), and contrasting this with human reasoning: we certainly have no list of all possible worlds in our brain, from which we choose the closest. Our reasoning is more flexible and constructive.

Here, composing situations might be the master logic, and remembering situations an auxiliary logic. The probably fundamental difference between logic and working of the brain will mostly be handled in the auxiliary logic. Before we take a more general look at the human brain and its functioning, we discuss very shortly the human memory, as it presents already many problems. (Note that, first, we are no neuroscientists, and, second, even neuroscientists have many conjectures, but much less established facts. This is, of course, due to the complexity of the subject.)

Human reasoning with counterfactual conditionals is much less regular than the formal approach in philosophical logic. It is not a procedure of simple choice by distance, but an active construction, a dialogue between different requirements. We might see it as a puzzle, where in addition, the tiles have to be cut to fit together.

3.3.1   Introduction

We discuss now our ideas how human beings think: not only in propositions and logical operators, or in models, but also in pictures, scenarios, prototypes, etc. On the neural level, such pictures correspond to groups of neurons.

We will be (necessarily) vague, on the meaning level, as well as the neural level. This vagueness results in flexibility, the price to pay are conceptual difficulties.

We have three types of objects:

(1) pictures or groups (of neurons),

(2) connections,

(3) attention.

Groups will be connected areas of the brain, corresponding to some picture, groups may connect to other groups with different types of connections, which may be positive or negative. Finally, attention focusses on groups or parts of groups, and their connections, or part thereof.

Groups do not necessarily correspond to nodes of graphs, as they are not atomic, and they can combine to new groups. Attention may hide contradictions, so the whole picture may contain contradictions.

We use the word "group" to designate

- on the physiological level a (perhaps only momentarily) somehow connected area of the brain, they may be formed and dissolved dynamically,

- on the meaning level a picture, scene (in the sense of conscious scene), a prototype (without all the connotations the word "prototype" might have), any fragment of information. It need not be complete with all important properties, birds which fly, etc., it might be a robin sitting on a branch in sunshine, just any bit of information, abstract, concrete, mixture of both, whatever.

3.3.2 Human Reasoning and Brain Structure

3.3.2.1 Pre-semantics and Semantics

In logic, a sentence like "it rains", or "if it were to rain, I would take an umbrella" has a semantics, which describes a corresponding state in the world.

We describe here what happens in our brain - according to our hypothesis - and what corresponds to this "brain state" in the real world. Thus, what we do is to describe an intermediate step between an expression in the language, and the semantics in the world.

For this reason, we call this intermediate step a pre-semantics.

In other words, real semantics interpret language and logic in (an abstraction of) the world. Pre-semantics is an abstraction of (the functioning of) the brain. Of course, the brain is "somehow" connected to the world, but this would then be a semantics of (the functioning of) the brain. Thus, this pre-semantics is an intermediate step between language and the world.

3.3.2.2 The Elements of Human Reasoning

In the following, we concentrate on episodes, pictures, etc., as they seem to be the natural structure to consider counterfactual conditionals. They also illustrate well the deep differences with the way formal logics work. For more details, see Section 3.5 (page 60) below.

It seems that humans (and probably other animals) often think in episodes, scenarios, prototypes, pictures, etc., which are connected by association, reasoning, developments, etc. We do not seem to think only in propositions, models, properties, with the help of logical operators, etc.

For simplicity, we call all above episodes, scenarios or pictures "pictures". Pictures can be complex, represent developments over time, can be combined, analysed, etc., they need not be precise, may be inconsistent, etc. We try to explain this, looking simultanously at the thoughts, "meanings" of the pictures, and the underlying neural structures and processes.

We imagine these scenarios etc. to be realised on the neural level by neurons or groups of neurons, and the connections between pictures also by neurons, or bundles of neurons.

To summarize, we have

(1) Pictures, on the

- meaning level, they correspond to thoughts, scenarios, pictures, etc.
- neural level, they correspond to neurons, clusters of connected neurons, etc.

(2) Connections or paths, on the

- meaning level, they correspond to associations, deductions, developments, coherences within an episode, etc.
- neural level, they correspond to neurons, bundles of more or less parallel neurons, connecting groups of neurons, etc.

All that follows is relative to this assumption about human reasoning.

Pictures: the Language and the Right Level of Abstraction

(1) The Level of Meaning

A picture can be complex, we can see it as composed of sub-pictures. A raven eats a piece of cheese, so there are a raven, a piece of cheese, perhaps some other objects in this picture. The raven has a beak, etc.

It is not clear where the "atomic" components are. On the neural level, we have single neurons, but they might not have meaning any more. In addition, they will usually not be accessible to conscious reasoning.

To solve this problem with atoms, we postulate that there are no atomic pictures, we can always decompose and analyse. For our purposes, this seems the best way out of the dilemma. In one context, the raven eating the cheese is the right level of abstraction, we might be interested in the behaviour of the raven. In a different context, it might be the feathers, the beak of the animal, or the taste of the cheese. Thus, there is no uniform adequate level of abstraction, it depends on the context.

(The "meaning" of a neuron, i.e. the conditions under which it fires, might be quite complicated. This is true even for relatively low-level neurons, i.e. close to sensory input: Visual information originating in the retina travels through the lateral geniculate nucleus in the thalamus to the visual cortex, first to V1. V1 itself is decomposed into 6 layers. Even cells in V1 receive feedback from higher-level areas like V4, which cover bigger and more complex receptive fields than those covered (directly) by V1 cells. This feedback can modify and shape responses of V1 cells. Thus, we may imagine such V1 cells to "say" something like: "I see an edge in my part of the retina, but context (sent by higher-level cells) thinks it is unlikely, still ...., etc." So, such cells may express rather complex situations. See [Wik17e], [Geg11], etc.

The organisation of V1 in "hypercolumns" going through the layers of V1 is very interesting, see the work by D. H. Hubel and T. Wiesel for which they were awarded a Nobel prize. (Roughly) edge detecting cells for one "spot" of the retina are grouped together, and moreoever, the direction of the edge which is detected changes continually. Thus, neighbouring "spots" have neighbouring groups of edge detectors, and, say, 0 degree detector sits close to 30 degree detector, 60 degree detector is farther away from the 0 degree detector, etc., in a cyclic way ("pinwheel"). This reminds of Hamming distances, where "spots" and degrees are the dimensions of the distance, and raises the question if other parts of the brain are organised similarly, with conceptually close information coded in neighbouring cells. The importance of neighbourhoods is seen, e.g., in the common-neighbor-rule (CNR), according to which a pair of neurons is more likely to be connected the more common neighbors it shares, see e.g. [AIZ16].)

(2) Relation to Models

Pictures will usually not be any models in the logical sense. There need not be any language defined, some parts may be complex and elaborate, some parts may be vague or uncertain, or only rough sketches, different qualities like visual, tactile, may be combined. Pictures may also be inconsistent.

(3) The Neural Level

On the neural level, pictures will usually be realised by groups of neurons, consisting perhaps of several thousands neurons. Those groups will have an internal coherence, e.g. by strong internal positive links among their neurons. But they will not necessarily have a "surface" like a cell wall to which other cells or viruses may attach. The links between groups go (basically) from all neurons of group 1 to all neurons of group 2. There is no exterior vs. interior, things are more flexible.

If group 1 "sees" (i.e. is positively connected to) all neurons of group 2, then group 2 is the right level of abstraction relative to group 1 (and its meaning). (In our example, group 1 looks at the behaviour of the raven.) If group 1a "sees" only a subgroup of group 2 (e.g. the feathers of the raven), either by having particularly strong connections to this subgroup, or by having negative connections to the rest of group 2, then this subgroup is the right level of abstraction relative to group 1a. Thus, the "right" level of abstraction is nothing mysterious, and does not depend on our speaking about the pictures, but is given by the activities of the neuron groups themselves.

(We neglect that changing groups of neurons may represent the same pictures.)

(4) The Conceptual Difficulties of this Idea

This description has certain conceptual difficulties.

(4.1) In logic, we have atoms (like propositional variables), from which we construct complex propositions with the use of operators like ∧, ∨, etc. Here, our description is "bottomless", we have no atoms, and can always look inside.

(4.2) There is no unique adequate level of abstraction to think about pictures. The right granularity depends on the context, it is dynamic.

(4.3) The right level of abstraction on the neural level is not given by our thoughts about pictures, but by the neural system itself. Other groups determine the right granularity.

(4.4) Groups of neurons have no surface like a cell or a virus do, there is no surface from which connections arise. Connections go from everywhere.

(5) Summary

(5.1) "Pictures" on the meaning level correspond to (coherent) groups of neurons.

(5.2) There are no minimal or atomic pictures and groups, they can always be decomposed (for our purposes). Single neurons might not have any meaning any more.

(5.3) Conversely, they can be composed to more complex pictures and groups.

(5.4) Groups of neurons have no surface, connections to other groups are from the interior.

3.3.2.3   Connections or Paths Between Neuron Groups

We sometimes call connections paths.

Connections may correspond to many different things on the meaning level. They may be:

(1) arbitrary associations, e.g. of things which happened at the same moment,

(2) inferences, classical or others,

(3) connections between related objects or properties, like between people and their ancestors, animals of the same kind, etc.,

(4) developments over time, etc.

Again, there are some conceptual problems involved.

(1) As for pictures, it seems often (but perhaps less dramatically) difficult to find atomic connections. If group N_1 is connected via path P to group N_2, but N_1' is a sub-group of N_1, connected via a subset P' of P to sub-group N_2' of N_2, then it may be reasonable to consider P' as a proper path itself.

(2) If, e.g., the picture describes a development over time, with single pictures at time t, t', etc. linked via paths expressing developments, then we have paths inside the picture, and the picture itself may be considered a path from beginning to end.

Thus, paths may be between pictures, or internal to pictures, and there is no fundamental distinction between paths and pictures. It depends on the context. More abstractly, the whole path is a picture, in more detail, we have paths between single pictures, "frames", as in a movie.

Remember: Everything is just suitably connected neurons!

3.3.2.4   Operations on Pictures and Connections

To simplify, we will pretend that operations are composed of cutting and composing. We are aware that this is probably artificial, and, more generally, an operation takes one or more pictures (on the meaning level) or groups (on the neural level), and constructs one or more new pictures or groups.

Before we describe our ideas, we discuss attention.

Attention

An additional ingredient is "attention". We picture attention as a light which shines on some areas of the brain, groups of neurons, perhaps only on parts of those areas, and their connections, or only parts of the connections. Attention allows, among other things, to construct a seemingly coherent picture by focussing only on parts of the picture, which are coherent. In particular, we might focus our attention on coherences, e.g., when we want to consolidate a theory, or incoherences, when we want to attack a theory. Focussing on coherences might hide serious flaws in a theory, or our thinking in general. In context A, we might focus on α, in context B, on β, etc. As we leave attention deliberately unregulated, changes in attention may have very "wild" consequences.

Activation means that the paths leading to the picture become more active, as well as the internal paths of the picture. Thus, whereas memory (recent use) automatically increases activity, attention is an active process. Conversely, pictures which are easily accessible (active paths going there), are more in the focus of our attention. E.g., we are hungry, think of a steak (associative memory), and focus our attention on the fridge where the steak is.

Attention originates in the "I" and its aims and desires. Likewise, "accessibility" is relative to the "I" - whatever that means. (This is probably a very simplistic picture, but suffices here. We conjecture that the "I" is an artifact, a dynamic construction, with no clear definition and boundary. The "I" might be just as elusive as atomic pictures.) Attention is related to our aims (find food, avoid dangers, etc.) and allows to focus on certain pictures (or parts of pictures) and paths.

Operations

Consider again the picture of a raven eating a piece of cheese.

We might focus our attention on the raven, and neglect the cheese. It is just a raven, eating something, or not. So the connection to the raven part will be stronger (positive), to the cheese part weaker positive and/or stronger negative. Conversely, we might never have seen a raven eat a piece of cheese. But we can imagine a raven, also a raven eating something, and a piece of cheese, and can put these pictures together. This may be more or less refined, adjusting the way the cheese lies on the ground, the raven pecks at it, etc. It is not guaranteed that the picture is consistent, and we might also adjust the picture "on the fly" to make it consistent or plausible. (When composing "raven" with "cheese" and "pecking", the order might be important: Composing "raven" first with "pecking" and the result "raven + pecking" with "cheese", or "raven" with "pecking + cheese" might a priori give different results.)

It is easy to compose a picture of an elephant with the picture of wings, and to imagine an elephant with small wings which hovers above the ground. Of course, we know that this is impossible under normal circumstances. There is no reality check in dreams, and a flying elephant is quite plausible.

This is all quite simple (in abstract terms), and everyone has done it. Details need to be filled in by experimental psychologists and neuroscientists. Obviously, these problems are related to planning.

Remarks on Composition

This might be a good place to elaborate our remarks of Section 3.2 (page 46) in the context of the functioning of the brain.

The working of the brain has often been described as "experts talking to each other", where the experts are different areas of the brain. So, we have a modular structure, and communication between the different modules. Why this structure? Is this only through evolution, which has added more structures to the brain, or can we see a different reason for this modularity? We think so.

Consider human experts discussing a situation, e.g. medical experts, a cardiologist, and a specialist for infectious disease, discussing a patient suffering from an unknown disease. They will discuss with each other every idea they have, but first try to come to some possible diagnoses, and then discuss the result of their thoughts. If one talks too early, the other might interrupt him: "let me think ....". Too much communication may disturb reflection. This is due to the fact that attention (here triggered by communication) might disturb a reflection process. More complicated reflection might involve considering rare situations, whose "signal" is weak, and this weak signal may easily be drowned by "loud" signals, e.g. from communication. Converse processes are concentration, attention, etc.

We have here an auxiliary process: active search for arguments, situations to consider. Both the reasoning itself, and the auxiliary process may be disturbed by "ouside noise". Thus, temporary isolation of reasoning modules may be helpful. The price to pay for the flexibility of the brain, shifting attention, various influences, is a necessary control over our reasoning processes. It seems complicated to cover all these aspects in a single logic, especially as many aspects will be dynamic, and there will be different processes running in parallel, with differing attentions, etc. Little seems to be known about control processes in human reasoning. A very interesting aspect is discussed in [WSFR02]. It is argued there that processing fluency is hedonically marked. Fluid processing elicits positive affective responses, visible in increased activity of the "smiling muscle". This would be a very high level control mechanism. It seems reasonable to ask if this is related to the fact that e.g. physicists emphasize the beauty of "right" theories, and consider the esthetic quality of a theory as an indication of its correctness. (The theory "flows".) If we want to model human reasoning, we will have to model such auxiliary processes, attention, etc., too.

Comparison to Operations in Logic

Consider an implication $A \to B$ in logic. To apply it, we have to look for A. Consider a picture $\alpha \to \beta$, which allows to be completed with α. But often, this will not be that precise. We can perhaps complete with some α', as there is no unique place to "dock" something. Perhaps we can complete with α, α', α'', etc., recall also that there is no surface to present all the docking possibilities, and no atoms in a useful sense. Suppose we have a picture $\alpha \to \beta \to \gamma$ describing some 3 stage development. Perhaps a picture δ may dock to α, but later, we see that it does not fit with β. So, we have local coherence, but not global coherence. A simplified analogon is perhaps putting together a puzzle.

We conjecture that we have active search for fitting pictures, composition itself, and then evaluation, i.e. checking for global fit. This seems quite complicated and it seems difficult to find general principles, analogue to logical properties, which govern such processes. (Language allows a different kind of flexibilty, based on categories of words, irrespective of their meaning.)

3.3.3 Counterfactual Conditionals

3.3.3.1 The Stalnaker/Lewis Semantics For Counterfactual Conditionals

Stalnaker and Lewis, see e.g. [Sta68], [Lew73], gave a very elegant semantics to counterfactual conditionals, based on minimal change. To give meaning to the sentence "if it were to rain, I would take an umbrella", we look at all situations (models) where it rains, and which are minimally different from the present situation. If I take an umbrella in all those situations, then the sentence is true. E.g., situations where there is hurricane - and I will therefore not take an umbrella - will, usually, be very different from the present situation.

This idea is very nice, but we do not think this way. First, we have no catalogue of all possible worlds in out head. Thus, we will have to compose the situations to consider from various fragments. Second, classical reasoning, taken for granted in usual semantics, has an "inference cost". E.g., when reasoning about birds, we might know that penguins are birds, but they might be too "far fetched", and forgotten.

It seems that human beings reason in pictures, scenes, perhaps prototypes, but in relatively vague terms. We try to use a more plausible model of this reasoning, based on neural systems, to explain counterfactual conditionals. But the basic Stalnaker/Lewis idea is upheld.

Our ideas, developped in Section 3.3.2 (page 49) are very rudimentary, all details are left open. Still, we think that it is a reasonable start. We might be overly flexible in our concepts, but it is probably easier to become more rigid later, than inversely.

We did not discuss how the various choices are made between different possibilities. On the one side, an overly rigid attention or memory might prevent flexibility, on the other side, too much flexibility might result in chaos and not enough focus. The brain "needs to roam", but with a purpose.

The Stalnaker/Lewis semantics is a passive procedure, there is a list of models, and we chose the "best" or "closest" with a certain property. Distance is supposed to be given. Our pre-semantics is much more active, we construct, disassemble, chose using several criteria. Thus, it is not surprising that control of the procedure is complex (and not discussed here).

3.3.3.2   The Umbrella Scenario

We apply our ideas to counterfactuals.

Note that the Stalnaker/Lewis semantics hides all problems in the adequate notion of distance, so we should not expect miracles from our approach.

"If it were to rain, I would take an umbrella."

We have the following present situations, where the sentence is uttered.

(1) Case 1: The "normal" case. No strong wind, I have at least one hand free to hold an umbrella, I do not want to get wet, etc.

(2) Case 2: As case 1, but strong wind.

(3) Case 3: As case 1, but I carry things, and cannot hold an umbrella.

We have the following pictures in our memory:

(1) Picture 1: Normal weather, it rains, we have our hands free, but forgot the umbrella, and get soaked.

(2) Picture 2: As picture 1, but have umbrella, stay dry.

(3) Picture 3: Rain, strong wind, use umbrella, umbrella is torn.

(4) Picture 4: As picture 1, but carry things, cannot hold umbrella, get soaked.

(5) Picture 5: The raven eating a piece of cheese.

Much background knowledge goes into our treatment of counterfactuals. For instance, that a strong wind might destroy an umbrella (and that the destruction of an umbrella in picture 3 is not due to some irrelevant aspect), that we need at least one hand free to hold an umbrella, that we want to stay dry, that we cannot change the weather, etc.

First, we actively (using attention) look for pictures which have something to do with umbrellas. Thus, in all cases, picture 5 is excluded.

Next, we look at pictures which support using an umbrella, and those which argue against this. This seems an enormous amount of work, but our experience tells us that a small number of scenarios usually give the answer. There are already strong links to those scenarios.

Case 1: Pictures 3 and 4 do not apply - they are too distant in the Stalnaker/Lewis terminology. So we are left with pictures 1 and 2. As we want to stay dry, we choose picture 2. Now, we combine case 1 with picture 2 by suitable connections, and "see" the imagined picture where we use an umbrella and stay dry.

Case 2: Pictures 1 and 4 do not apply. I would prefer to stay dry, but a torn umbrella does not help. In addition, I do not want my umbrella to be torn. Combining case 2 with picture 3 shows that the umbrella is useless, so I do not take the umbrella.

Case 3: Only picture 4 fits, I combine and see that I will get wet, but there is nothing I can do.

3.3.3.3 A Tree Felling Scenario

Consider the sentence:

"If I want to fell that tree, I would hammer a pole into the ground, and tie a rope between tree and pole, so the tree cannot fall on the house."

(This is an anankastic conditional, see e.g. [Sab14], but the lingistic problems need not bother us. By the way, the following statement: "If you want to jump to the moon, you should wait for a clear night with full moon, so you do not miss it." might be fun to look at.)

We have the present situation where the tree stands close to the house, there are neither rope nor pole, nor another solid tree where we could anchor the rope, and we do not want to fell the tree.

We have

- Picture 1 of a pole being hammered into the ground - for instance, we remember this from camping holidays.

- Picture 2 of a rope tied to a tree and its effect - for instance, we once fastened a hammock between two young trees and saw the effect, bending the trees over.

- Picture 3 of someone pulling with a rope on a big tree - it did not move.

We understand that we need a sufficiently strong force to prevent the tree from falling on the house.

Pictures 2 and 3 tell us that a person pulling on the tree, or a rope tied to a small tree will not be sufficient.

As there is no other sturdy tree around, we have to build a complex picture. We have to cut up the hammock Picture 2 and the tent Picture 1, using the rope part from Picture 2, the pole part from Picture 1. It is important that the pictures are not atomic. Note that we can first compose the situation with the pole part, and the result with the rope part, or first the situation with rope part, and the result with the pole part, or first combine the pole part with the rope part, and then with the situation. It is NOT guaranteed that the outcome of the different ways will be the same. When there are more pictures to consider, even the choice of the pictures might depend on the sequence.

3.3.4   Comments

There are many aspects we did not treat. We established a framework only.

(1) Usually, there are many pictures to choose. How do we make the choice?

(2) How do we cut pictures?

(3) How do we determine if a combined picture is useful?

(4) Attention can hide inconsistencies, or focus on inconsistencies, how do we decide?

(5) Are all these processes on one level, or is it an interplay between different levels (execution and control)?

(6) These processes seem arbitrary, but we are quite successful, so there must be a robust procedure to find answers.

Some of the answers will lie in the interplay between (active) attention and more passive memory (more recent and more frequently used pictures and processes are easier accessible). Recall here Edelman's insight, see e.g. [Ede89], [Ede04], that there are parallels between the brain and the immune system, both working with selection from many possibilities. We assume that we have many candidates of the same type, so we have a population from which to chose. We chose the best, and consider this set for the properties of those combined areas.

It is natural to combine the ideas of the hierarchy in [GS16], chapter 11 there, with our present ideas. Exceptional classes, like penguins, are only loosely bound to regular classes, like birds; surprise cases even more loosely.

3.4   More Formal Remarks

We do not have elementary propositions, nor operators like $\wedge$ etc. Instead, we have (groups of) neurons, and connections between them. The possibility of "pruning", see below, captures the fact that pictures/situations are not elementary. Conversely, it is possible to combine pictures.

Attention (on the level of active search, pruning, combining) considers choice and operations on situations. On the level of evaluation, attention considers the result of the operations.

We give now a simple abstraction of networks of neurons. The basics are common knowledge, see e.g. [GLP17].

Definition 3.4.1

The body of a neuron is a counter, which counts positive (excitatory) inputs into the neuron, subtracts negative (inhibitory) inputs, if the result is above a certain (individual for this neuron, but static value) threshold, the neuron become active (fires), otherwise, it stays dormant.

Consequently, we have a simple property: If a neuron is active, increasing the positive input and/or decreasing the negative input will keep it active. Conversely, if a neuron is dormant, decreasing the positive input and/or increasing the negative input will keep it dormant.

Definition 3.4.2

Neuron bodies can be connected via arrows. Each arrow has an integer value, which may change over time. For simplicity, we assume that there is only one arrow between two neurons (per direction). (If there are several, we code this by modifying the value.)

Arrows are dynamic over time:

(1) New arrows may be created, and arrows may disappear.

(2) The values of arrows may change over time.

In particular:

(1) Recent use: Use of a positive connection between neuron a and neuron b may increase the value of the arrow $a \rightarrow b$, lack of use may decrease the value. This results in stronger bonds for frequent situations, but also a certain rigidity.

(2) Hebb's rule, see Definition 3.5.1 (page 61): If neurons a and b are activated at the same time, e.g. by positive arrows $c \rightarrow a$ and $c \rightarrow b$, then the connection between a and b is strengthened (i.e. arrows $a \rightarrow b$ and $b \rightarrow a$).

(3) Attention: Attention is a more active way of modifying arrows, but like past use, it may modify the values of arrows in both directions (and also create

new ones). See e.g. [Auf17]. (Despite the word "attention", we do not think that consciousness is a necessary condition for attention. Even simple animals need to focus, to escape from predators over search for food, etc.)

More aspects:

(1) Active search: Attention can focus on some aspects, e.g. rain, umbrella in our example, of a picture, and search for other pictures with the rain/umbrella element. A positive connection is made to such situations.

(2) Pruning: Attention may neglect some aspects of a situation, and make e.g. negative connections to those aspects.

(3) Combining: Attention may combine two or more situations by making positive connections between them, making them components of a more complex situation.

(4) Evaluation: The result of these operations may be evaluated again. The criteria will be (in an incomplete list)

- are important parts of the situation we found neglected (e.g., there was a strong wind, which destroyed the umbrella)?
- is the constructed situation sufficiently coherent?
- does it seem necessary to search for competing situations?

Evaluation may lead to backtracking, new attempts, etc.

3.5   Appendix - Some Remarks on Neurophilosophy

3.5.1   Introduction

We give here a very short summary of aspects of human reasoning which are important in our context. Human reasoning and the functioning of the brain are extremely complex, and largely still unknown, we only indicate roughly some aspects. (There is a vast literature, and we just mention some we looked into: [Sta17d], [Rot96], [Chu89], [Chu86], [Chu07], [Wik17a], [Wik17b], [Wik17c], [Wik17d], [CCOM08], [HM17], [KPP07], [ZMM15], [Wik17e], [Geg11], [AIZ16], [Pul13].)

We think some - even rudimentary understanding - is important in our context for the following reason: The more a logic might seem close to human reasoning, the more differences to human reasoning might be important. For instance, the Stalnaker/Lewis semantics for counterfactual conditionals (CFC's), see [Sta68], [Lew73], is intuitively very attractive, so we might be tempted to see it as describing actual human reasoning. This may have serious consequences. In court, a defendant might say "if I had done .... then ...." in good faith, following his own reasoning, but a judge familiar with the theory of formal CFC's might come to a different conclusion and accuse the defendant of lying. See e.g. [Wik16a] and [IEP16] for different legal systems. We will come back to CFC's below.

Just as classical logic is not a description of actual human reasoning, concepts of philosophical logic need not correspond directly to the way we think.

Thus, our remarks are also a warning against hasty conclusions. But, of course, we may speculate on the (neural) naturalness of concepts like "distance", which have an analogon in the brain, the strength of the connection between areas of the brain, or between groups of neurons.

The author is absolutely no expert on neuroscience. In addition, it seems that recent research has concentrated on the structure and function of single or small numbers of nerve cells, and somewhat neglected the overall picture of how our brain works. Thus, there does not seem to exist an abstract summary of present knowledge about human reasoning on the neural level. Perhaps, one should more dare to be wrong, but incite criticism, and thus advance our knowledge?

3.5.2   Details

3.5.2.1   Basics

The basic unity of a nervous system are neurons, consisting of dendrites, core, and axon. Signals travel from the axon of neuron A via a synapse, the connection, to a dendrite of neuron B, etc. Usually, a neuron has one axon and several dendrites. The axon of cell A may be connected via synapses to the dendrites of several neurons B, B', etc.

A synapse may be positive or negative, excitatory or inhibitory. Suppose neuron B is in exited level b, and it receives a signal from neuron A via an excitatory synapse, then B goes to level $b' > b$; if the synapse is inhibitory, it goes to $b' < b$. As B might receive several signals, (very roughly) the sum of incoming signals, positive or negative, determines b'-b.

More precisely, if, within a certain time interval, the sum of positive signals Σ^+, i.e. from positive synapses, arriving at the dendrites of a given neuron is sufficiently bigger than the sum of negative signals Σ^-, i.e. from negative synapses, arriving at the dendrites of the same neuron, the neuron is activated and will fire, i.e. send a signal via its axon to other neurons. This is a 0/1 reaction, it will fire or not, and always with the same strength. (If Σ^+ is much bigger than Σ^-, the neuron may fire with a higher frequency. We neglect this here.)

Note that negative synapses are "related" to negation, but are not negation in the usual sense (nor negative arrows in defeasible inheritance systems). They rather express (roughly) "contradict each other" like "black" and "yellow" do.

An extremely important fact is Hebb's rule (see [Heb49]):

Definition 3.5.1

When neurons A and B are simultaneously activated, and are connected via some synapse, say from A to B, then the connection is strenthened, i.e. the weight of the synapse is increased, and thus the future influence of A on B is increased. This property is also expressed by: "fire together, wire together". The dynamic history is thus remembered as an association between neurons (or groups of neurons).

3.5.2.2   The "Meaning" and Dynamics of the Activity of a Neuron

If a photoreceptor cell in the eye is excited (by light or pressure, etc.), it will always send the message "light". Hair cells in the ear detect sounds, they send the message "sound". These determine the different qualia, light and sound.

In general, things are not so simple. First, activities of the brain usually involve many, perhaps thounds of neurons, which work together as a (strongly interconnected) group of neurons. Second, this activity may involve in moment a a neuron group A, in moment b neuron group B, etc. (A good example is a cloud which "sits" seemingly stationary on top of a mountain in a strong wind. As a matter of fact, single water vapor molecules are pushed upwards over the mountain, they condense as they cool down, they reflect light, and become visible. When they descend again, they warm, evaporate, and "disappear" on the other side of the mountain. New water molecules follow, so the overall picture is static, the components which create the picture change all the time.) This also results in the extreme flexibility of the brain. Usually, the death of one neuron may be compensated by another neuron.

Thus, in general, the "meaning" of an active (i.e. firing or close to firing) neuron or a group of active neurons is defined by the context within the present active network of the brain. This is called the functional role semantics of neurons or groups of neurons. The state of the brain (activities, strengths of synapses, etc.) may be seen as an extremely complex vector, and its transformation from one state to the other as a vector transformation. Thus, the mathematics of dynamical systems seems a promising approach to brain activity.

Consider a picture or scene coded (at present, and static for simplicity) by the activity of some group of neurons. E.g., we look through the window an the garden. Then:

(1) There are no atoms, we can always analyze parts of the picture even further. Of course, we have neurons as "atoms" on the neural level, but they have lost their meaning, which exists only in the context. Thus, there are no atoms of thought.

(2) The group of neurons has no "surface". One group of neurons is not "seen" by another group of neurons with presenting some opaque surface, but, usually, connections between the two groups go as well between the "surfaces" - how ever they may be defined - as between the interior neurons.

There is no "right" level of abstraction or granularity to consider this picture - it depends on the connection to other groups of neurons. Fix a neuron group G. One (other) group of neurons may be connected to, "sees" all neurons of G, another group of neurons may be connected to, "sees" only a subset of G.

Suppose we see a raven eating a bit of cheese in the garden. If we are interested in ravens, we will focus our attention on different aspects of the scenario, than if we want to buy cheese, and this reminds us not to forget.

(3) In addition, the connections to other groups of neurons may themselves be complex, and may consist again of many neurons, they are not simple operators like $\wedge$, $\neg$ etc. as between words in a language. So, groups of neurons are connected to other groups of neurons via groups of neurons, and the same considerations as above apply to the connecting groups.

(3.1) The use of a group of neurons makes this group easier accessible, and strengthens its internal coherence. Thus, the normal case becomes stronger.

(3.2) The use of a neural connection strengthens this connection. Again, the normal connections become stronger. Both properties favour learning, but may also lead to overly rigid thinking and prejudice. Note that this is the opposite of basic linear logic, where the use of an argument may consume it.

(3.3) When two groups of neurons, N_1 and N_2 are activated together, this strengthens the connection between N_1 and N_2 by Hebb's rule, see Definition 3.5.1 (page 61), and e.g. [Pul13]. This property establishes associations. When I hear a roar in the jungle, and see an attacking tiger, next time, I will think "tiger" when I hear a roar, even without seeing the tiger.

(3.4) A longer connection may be weaker. For instance, penguins are an abnormal subclass of birds. Going from birds to penguins will not be via a strong connection (though *penguin* $\to$ *bird* is a classical inference). If Tweety is a penguin, we might access Tweety only by detour through penguin. Thus, Tweety is "less" bird than the raven which I saw in my garden. Consequently, the subset relation involved in the properties of many nonclassical logics has a certain "cost", and the resulting properties (e.g. $X \subseteq Y \to \mu(Y) \cap X \subseteq \mu(X)$ for basic preferential logic, similar properties for theory revision, update, counterfactual conditionals) cannot always be expected.

We summarize:

- Connections are made of single axon-synapse-dendrite tripels, connecting one neuron to another, or many such tripels, bundels, or composed bundels. Again, it seems useful to say that connections can usually be decomposed into sub-connections.
- Connections can be via excitatory or inhibitory synapses, the former activate the downstream neuron, the latter de-activate the downstream neuron.
- Connections can have very different meanings.
- Connections can be interior to groups, or between groups.

- There usually is no clear distinction between groups and connections.

(4) It is important to note fundamental differences to (formal) languages:

There is no "right" level of abstraction or granularity to consider this picture - it depends on the connection to other groups of neurons.

The connections to other groups of neurons are themselves complex, and may consist again of many neurons, they are not simple operators like $\land$, $\neg$ etc.

3.5.2.3   Organisation of the Brain

(1) Recursiveness

The processing of information via connections is not linear, but recursive, even in relatively basic (visual) circuits.

Consequently, there must be some mechanism preventing wild oscillations and uncontrolled reinforcement, see also Chapter 6 (page 105).

Thus, a hypothesis might enhance lower level attention to certain aspects, and help decide about truth of the hypothesis, which, by its organisation, the lower level center might be unable to do on its own.

(2) Different areas of the brain

There are different areas for different tasks of the brain. E.g., there are areas for language processing, and there is a semantic memory for facts and concepts, an episodic memory for events, experiences, scenes, etc. The latter allows to construct new scenes from old ones, etc. Scenes will be memorised by (intraconnected) groups of active neurons, consisting of perhaps thousands of neurons.

The brain has grown in the course of evolution, but it is not an "organic" growth, it is rather like a shanty town, where old parts are still being used, or cooperating with newer parts.

Most important, the brain has to be sufficiently connected to the (natural and social) world to be useful.

(3) There are different types of memory in the human brain.

A look at the literature on human memory, e.g., ([Wik17a], [Wik17b], [Wik17c], [Wik17d], [CCOM08], [HM17], [KPP07], [ZMM15]) shows that:

(3.1) There are several memories, and some tasks and problems due to lesions allow to differentiate between different memories.

There are e.g. short-term and long-term memory, within long-term memory implicit memory (unconcious, concerning skills etc.) vs explicit memory (conscious, declarative), within declarative memory episodic memory (events, experiences, scenarios, pictures) vs. semantic memory (facts, concepts), etc.

We are interested here in episodic and semantic memory. Episodic memory concerns episodes within a context, "stories", semantic memory concerns principles and facts independent of context.

As the word "semantics" is heavily used in logics, we will call semantic memory "conceptual memory" here.

Episodic memory consists of (intraconnected) groups of active neurons, of perhaps thousands of neurons.

(3.2) If episodic memory were based on conceptual memory (a number of connected conceptual entities over time), or vice versa (the common feature of a number of episodes), then failure of one system would also cause failure of the other system. This is not the case. On the other hand, it seems unlikely that both systems coexist without connections. The exact connections and independencies seem unclear. Of course, this is a fundamental problem, and extremely important to the relation between the world, our brain, and our language.

(3.3) There seem to be memory structures in the strict sense (where information is stored), and auxiliary structures, e.g. for storing (writing) and accessing (reading) information. But this separation might not be strict. Acessing memory may be top-down (e.g. "where did I park the car?") or bottom-up (e.g. image of the car near a tree).

(3.4) Memory activities might involve several centers of the brain, e.g. visual memory might involve the centers for processing visual information, likewise for auditory memory, etc. These questions are not settled.

(3.5) There are several models of conceptual memory, e.g. network models, and feature models. The first are types of neuronal networks, the latter close to defeasible inheritance networks. - The location of the conceptual memory in the brain is not clear. It might be a collection of functionally and anatomically distinct systems.

(3.6) Episodic memory allows to construct new scenes from past scenes, insertions, blending, cutting, etc. It is not clear where this happens, in the episodic memory itself, or in a "higher" structure, or both? Is there a control mechanism which supervises the result (and is perhaps asleep when we dream)? - The parietal cortex and the hippocampus seem to be involved in episodic memory.

(3.7) Learning is believed to be a Hebbian process, see Definition 3.5.1 (page 61).

(4) The brain does not contain 1-1 images of the world

Neither the present image of the world nor the memory is a simple 1-1 image. In both cases, past experiences, desires, attention, etc. form and deform the image.

We cannot exclude that there are aspects of the world which are, in principle, inaccessible to our thoughts. Our existence proves that our brain is sufficiently adapted to the world to deal with it. It seems difficult to go beyond this scepticism.

3.5.2.4   The (limited) Role of Language

(1) Obviously, not all knowledge is coded by language. Sexing (determining the sex of) chicken is a famous example. Experts (here in sexing) may be very efficient, but unable to express their knowledge by words.

(2) Even scientific theories seem to have non-verbal aspects.

(3) In legal reasoning, even fundamental ideas may be expressed by words with unclear meaning. A famous example is the German constitution which has as basic concept "dignity of man". This concept has a long history, but no clear meaning. At first sight, it seems absurd to base a constitution on an unclear concept, leaving a wide margin for interpretation, but the authors of the constitution may have been confident that judges and law makers will know how to apply it (as experts know how to sexe chicken). Thus, "dignity of man" is perhaps best seen as a pointer (or label) to ways to interpret it, as a collection of prototypes.

(4) The (philosophical) notion of an ideal might be seen in the same way (and not as a list of properties which fails, as is well known).

(5) A more general problem is to describe brain states which are principally unreachable by language - if they exist. (And, even more generally, are there brainstates A and B, which have only limited communication possibilities?)

(6) Language may help to structure, stabilise, and refine knowledge: wine connaisseurs use seemingly bizarre expressions to describe their experiences.

(7) Expressions of our languages have no direct semantics in the world, but first in brain states, and those brain states somehow correspond to the world. Thus the first meaning of expressions is a kind of "pre-semantics" (not between formal language and formal semantics, but a brain activity between natural language and the "real world").

3.5.2.5   Comparison to Defeasible Inheritance Networks

For an overview of defeasible inheritance, see e.g. [Sch97-2].

Note that several neurons may act together as an amplifier for a neuron: Say N_1 connects positively to N_2 and N_3, and N_2 and N_3 each connect positively to N_4, and N_1 is the only one to connect to N_2 and N_3, then any signal from N_1 will be doubled in strength when arriving at N_4, in comparison to a direct signal from N_1 to N_4.

(1) Consequently, direct links do not necessarily win over indirect paths. If, in addition, N_1 is connected negatively to N_4, then the signal from N_1 to N_4 is 2 for positive value (indirect via N_2 and N_3) and -1 for negative value (direct to N_4), so the indirect paths win.

(2) By the same argument, longer paths may be better.

On the other hand, a longer path has more possibilities of interference by other signals of opposite polarity.

So length of path is no general criterion, contrary to inheritance systems, where connections correspond to "soft" inclusions.

(3) The number of paths of the same polarity is important, in inheritance systems, it is only the existence.

(4) There is no specificity criterion, and no preclusion.

(5) In inheritance systems, a negative arrow may only be at the end of a path, it cannot continue through a negative arrow. Systems of neurons are similar: If a negative signal has any effect, it prevents the receiving neuron from firing, so this signal path is interrupted.

(6) Neurons act directly sceptically - there is no branching into different extensions.

3.6   Acknowledgements

The author is much indebted to C.v.d. Malsburg, FIAS, Frankfurt, for patient advice, and to D. Gabbay for discussions.

Chapter 4

A Comment on Analogical Reasoning

4.1 Introduction

4.1.1 Overview

Our idea originated from the remarks of Section 2.4 of [SEP13] and the scepticism expressed there to find a logic for analogical reasoning, see Section 4.1.3 (page 71) below.

In a way, our (positive) reply is a form of cheating: we avoid the problem, and push it and the solution into a suitable choice function - as is done for Counterfactual Conditionals, Preferential Structures, etc. We then have a characterisation machinery we can just take off the shelf, and we have results for ANY choice function.

We present in the rest of this introduction (largely verbatim, only punctually slightly modified) excerpts from [SEP13], see also [SEP19c], to set the stage for the answer in Section 4.2 (page 72).

4.1.2 Section 2.2 of [SEP13]

(These remarks concern p. 5-7 of Section 2.2 of [SEP13].)

Definition 4.1.1

An analogical argument has the following form:

1. S is similar to T in certain (known) respects.

2. S has some further feature Q.

3. Therefore, T also has the feature Q, or some feature $Q*$ similar to Q.

(1) and (2) are premises. (3) is the conclusion of the argument. The argument form is inductive in the sense that the conclusion is not guaranteed to follow from the premises.

S and T are referred to as the source domain and target domain, respectively. A domain is a set of objects, properties, relations and functions, together with a set of accepted statements about those objects, properties, relations and functions. More formally, a domain consists of a set of objects and an interpreted set of statements about them. The statements need not belong to a first-order language, but to keep things simple, any formalizations employed here will be first-order. We use unstarred symbols (a, P, R, f) to refer to items in the source domain and starred symbols ($a*$, $P*$, $R*$, $f*$) to refer to corresponding items in the target domain.

Definition 4.1.2

Formally, an analogy between S and T is a one-to-one mapping between objects, properties, relations and functions in S and those in T.

J. M. Keynes, in [Key21], introduced some helpful terminology:

(1) Positive analogy.

Let P stand for a list of accepted propositions $P_1, \ldots, P_n$ about the source domain S. Suppose that the corresponding propositions $P_1* , \ldots, P_n*$, abbreviated as $P*$, are all accepted as holding for the target domain T, so that P and $P*$ represent accepted (or known) similarities. Then we refer to P as the positive analogy.

(2) Negative analogy.

Let A stand for a list of propositions $A_1, \ldots, A_r$ accepted as holding in S, and $B*$ for a list $B_1*, \ldots, B_s*$ of propositions holding in T. Suppose that the analogous propositions $A* = A_1*, \ldots, A_r*$ fail to hold in T, and similarly the propositions $B = B_1, \ldots, B_s$ fail to hold in S, so that A, $\neg A*$ and $\neg B$, $B*$ represent accepted (or known) differences. Then we refer to A and B as the negative analogy.

(3) Neutral analogy.

The neutral analogy consists of accepted propositions about S for which it is not known whether an analogue holds in T.

(4) Hypothetical analogy.

The hypothetical analogy is simply the proposition Q in the neutral analogy that is the focus of our attention.

These concepts allow us to provide a characterization for an individual analogical argument that is somewhat richer than the original one.

Definition 4.1.3

(Augmented representation)

Correspondence between SOURCE (S) and TARGET (T)

(1) Positive analogy:

$P \Leftrightarrow P*$

(2) Negative analogy:

$A \Leftrightarrow \neg A*$

and

$\neg B \Leftrightarrow B*$

(3) Plausible inference:

$Q \Leftrightarrow Q*$

An analogical argument may thus be summarized: It is plausible that $Q*$ holds in the target because of certain known (or accepted) similarities with the source domain, despite certain known (or accepted) differences.

4.1.3   Section 2.4 of [SEP13]

Scepticism:

Of course, it is difficult to show that no successful analogical inference rule will ever be proposed. But consider the following candidate, formulated using the concepts of the schema in Definition 4.1.3 (page 71) and taking us only a short step beyond that basic characterization.

Definition 4.1.4

Suppose S and T are the source and target domains. Suppose $P_1, \ldots, P_n$ represents the positive analogy, $A_1, \ldots, A_r$ and $\neg B_1, \ldots, \neg B_s$ represent the (possibly vacuous) negative analogy, and Q represents the hypothetical analogy. In the absence of reasons for thinking otherwise, infer that $Q*$ holds in the target domain with degree of support $p > 0$, where p is an increasing function of n and a decreasing function of r and s.

(Definition 4.1.4 (page 71) is modeled on the straight rule for enumerative induction and inspired by Mill's view of analogical inference, as described in [SEP13] above. We use the generic phrase "degree of support" in place of probability, since other factors besides the analogical argument may influence our probability assignment for $Q*$.)

The schema in Definition 4.1.4 (page 71) justifies too much.

So, how do we chose the "right one"?

Remark 4.1.1

The author was surprised to find a precursor to his concept of homogenousness, see Chapter 5 (page 77), and [Sch97-2], [GS16], in the work of J. M. Keynes, [Key21], quoted in Section 4.3 of [SEP13].

4.2   The Idea

We now describe the idea, and compare it to other ideas in philosophical and AI related logics.

But first, we formalize above ideas into a definition.

Definition 4.2.1

Let $\mathcal{L}$ be an alphabet.

(1) Let $\mathcal{L}_\alpha \subseteq \mathcal{L}$, and $\alpha : \mathcal{L}_\alpha \to \mathcal{L}$ an injective function, preserving the type of symbol, e.g.,

- if $x \in \mathcal{L}_\alpha$ stands for an object of the universe, then so will $\alpha(x)$
- if $X \in \mathcal{L}_\alpha$ stands for a subset of the universe, then so will $\alpha(X)$
- if $P(.) \in \mathcal{L}_\alpha$ stands for an unary predicate of the universe, then so will $\alpha(P)(.)$
- etc., also for higher symbols, like $f : \mathcal{P}(U) \to \mathcal{P}(U)$, U the universe.

(2) Let $\mathcal{F}_\alpha$ a subset of the formulas formed with symbols from $\mathcal{L}_\alpha$.

For $\phi \in \mathcal{F}_\alpha$, let $\alpha(\phi)$ be the obvious formula constructed from ϕ with the function α.

(3) We now look at the truth values of ϕ and $\alpha(\phi)$, $v(\phi)$ and $v(\alpha(\phi))$. In particular, there may be ϕ s.t. $v(\phi)$ is known, $v(\alpha(\phi))$ not, and we extrapolate that $v(\phi) = v(\alpha(\phi))$, this is then the analogical reasoning based on α.

More precisely:

(3.1) There may be ϕ s.t. $v(\phi)$ is not known, $v(\alpha(\phi))$ is known or not, such ϕ do not interest us here.

Assume in the following that $v(\phi)$ is known.

(3.2) $v(\phi)$ and $v(\alpha(\phi))$ are known, and $v(\phi) = v(\alpha(\phi))$. The set of such ϕ is the positive support of α, denoted α^+.

(3.3) $v(\phi)$ and $v(\alpha(\phi))$ are known, and $v(\phi) \neq v(\alpha(\phi))$. The set of such ϕ is the negative support of α, denoted α^-.

(3.4) $v(\phi)$ is known, $v(\alpha(\phi))$ is not known. The set of such ϕ is denoted $\alpha^?$. The "effect" of α is to conjecture, by analogy, that $v(\phi) = v(\alpha(\phi))$ for such ϕ.

Intuitively, α^+ strengthens the case of α, α^- weakens it - but these need not be the only criteria, see also [SEP13] and [SEP19c].

Such α - or a modification thereof - are the basic concepts of analogy. Source and destination of α often describe aspects of the "world", α itself need not correspond to anything in the world (like a common cause), but may be merely descriptional. Moreover, there may also be "meta-analogies" between analogies.

Let $\mathcal{A}$ be a set of functions α as defined in Definition 4.2.1 (page 72).

We may close $\mathcal{A}$ under combinations, as illustrated in the following Example 4.2.1 (page 73), or not.

Example 4.2.1

Consider α, α'.

Let $x, x', P, Q \in \mathcal{L}_\alpha = \mathcal{L}_{\alpha'}$, $Q(x), Q(x') \in \alpha^?, \alpha'^?$.

(1) α works well for x, but not for x' : $P(x) = \alpha(P)(x)$, $P(x') \neq \alpha(P)(x')$, so $P(x) \in \alpha^+$, $P(x') \in \alpha^-$,

(2) α' works well for x', but not for x : $P(x') = \alpha'(P)(x')$, $P(x) \neq \alpha'(P)(x)$, so $P(x) \in \alpha'^-$, $P(x') \in \alpha'^+$.

Let further $\alpha(Q)(x) \neq \alpha'(Q)(x)$ and $\alpha(Q)(x') \neq \alpha'(Q)(x')$.

What shall we do, should we chose one, α or α', for guessing, or combine α and α' to α'', chosing $\alpha'' = \alpha$ for expressions with x, and $\alpha'' = \alpha'$ for expressions with x', more precisely $\alpha''(Q)(x) := \alpha(Q)(x)$, and $\alpha''(Q)(x') := \alpha'(Q)(x')$?

The idea is now to push the choice of suitable $\alpha \in \mathcal{A}$ into a relation $\prec$, expressing quality of the analogy. E.g., in Example 4.2.1 (page 73), $\alpha'' \prec \alpha$ and $\alpha'' \prec \alpha'$ - for historical reasons, smaller elements will be "better".

Usually, this "best" relation will be partial only, and there will be many "best" f. Thus, it seems natural to conclude the properties which hold in ALL best f.

Definition 4.2.2

Let $\mathcal{A}$ be a set of functions as described in Definition 4.2.1 (page 72), and $\prec$ a relation on $\mathcal{A}$ (expressing "better" analogy wrt. the problem at hand).

We then write $\mathcal{A} \models_\prec \phi$ iff ϕ holds in all $\prec -$best $\alpha \in \mathcal{A}$.

(This is a sketch only, details have to be filled in according to the situation considered.)

4.2.1   Discussion

This sounds like cheating: we changed the level of abstraction, and packed the question of "good" analogies into the $\prec$-relation.

But when we look at the Stalnaker-Lewis semantics of counterfactual conditionals, see [Sta68], [Lew73], the preferential semantics for non-monotonic reasoning and deontic logic, see e.g. [Han69], [KLM90], [Sch04], [Sch18], the distance semantics for theory revision, see e.g. [LMS01], [Sch04], this is a well used "trick" we need not be ashamed of.

In above examples, the comparison was between possible worlds, here it is between usually more complicated structures (functions), yet this is no fundamental difference.

But even if we think that there is a element of cheating in our idea, we win something: properties which hold in ALL preferential structures, and which may be stronger for stronger relations $\prec$, see Fact 4.2.1 (page 74) below.

4.2.2 Problems and Solutions

(1) In the case of infinitely many α's we might have a definability problem, as the resulting best guess might not be definable any more - as in the case of preferential structures, see e.g. [Sch04].

(2) Abstract treatment of representation problems for abovementioned logics works with arbitrary sets, so we have a well studied machinery for representation results for various types of relations of "better" analogies - see e.g. [LMS01], [Sch04], [Sch18].

To give the reader an idea of such representation results, we mention some, slightly simplified (we neglect the multitude of copies for simpler presentation).

Definition 4.2.3

(1) Let again $\prec$ be the relation, and $\mu(X) := \{x \in X : \neg \exists x' \in X. x' \prec x\}$,

(2) $\prec$ is called smooth iff for all $x \in X$, either $x \in \mu(X)$ or there is $x' \in \mu(X)$, $x' \prec x$,

(3) $\prec$ is called ranked iff for all x, y, z, if neither $x \prec y$ nor $y \prec x$, then if $z \prec x$, then $z \prec y$, too, and, analogously, if $x \prec z$, then $y \prec z$, too.

We then have e.g.

Fact 4.2.1

(2.1) General and transitive relations are characterised by

$(\mu \subseteq) \ \mu(X) \subseteq X$

and

$(\mu PR) \ X \subseteq Y \rightarrow \mu(Y) \cap X \subseteq \mu(X)$

(2.2) Smooth and transitive smooth relations are characterised by $(\mu \subseteq)$, (μPR), and the additional property

$(\mu CUM) \ \mu(X) \subseteq Y \subseteq X \rightarrow \mu(X) = \mu(Y)$

(2.3) Ranked relations are characterised by $(\mu \subseteq)$, (μPR), and the additional property

$(\mu =)\ X \subseteq Y,\ \mu(Y) \cap X \neq \emptyset \rightarrow \mu(X) = \mu(Y) \cap X.$

For more explanation and details, see e.g. [Sch18], in particular Table 1.6 there.

Chapter 5

Relevance and Homogenousness

5.1 Introduction

Erledigt bis 5.2.2 einschliesslich am 21.4.22

We keep this introduction short and refer the reader for more details to [GS16], Chapter 11, and [Sch18b], Chapter 5.

5.1.1 The Principle of Homogenousness

Definition 5.1.1

The principle of homogenousness can be stated, informally, as:

"Given an set X of cases, and $X' \subseteq X$, the elements of X' usually behave as the elements of X do."

The strict version (without the "usually") is obviously wrong, but the soft version (as above) is an extremely useful hypothesis, and it seems impossible for living beings even to survive without it.

The principle of homogenousness is implicit in many systems of NML.

It seems impossible to understand the intuitive justification of the downward version in the default (in the sense of Reiter) and the defeasible inheritance variants of NML, without accepting a default (in the intuitive sense) version of homogenousness: If, in X, normally ϕ holds, why should ϕ normally hold in a subset X' of X (unless X' is "close to" X) - if we do not accept some form of homogenousness? Of course, by the very principle of non-monotonicity, this need not be the case for all $X' \subseteq X$, as X' might just be the set of exceptions. (See the discussion in [Sch97-2].)

We are mainly interested in non-monotonic logic, whose consequence relation is often written $\sim$, and, by their very nature, $X \sim \phi$ does not impl;y $X' \sim \phi$ - though it will "often" hold. Thus, for some properties, the "border" between X and X' is relevant, for many it is not.

Depending on the strength of the underlying non-monotonic logic $\sim$, we have "hard" second order properties, like: if X differs little in size from Y, and $X \sim \phi$, then $Y \sim \phi$. We want to go beyond these hard second order properties, and look at reasonable other second order properties and their relation to each other. We will argue semantically, and abstract size will be central to our work. Many properties will be default coherence properties between X, Y, etc.

We have now the following levels of reasoning:

(1) Classical logic:

   monotony, no exceptions, clear semantics

(2) Preferential logic:

   small sets of exceptions possible, clear semantics, strict rules about exceptions, like (μCUM), no other restrictions

(3) Meta-Default rules (Homogenousness):

   They have the form: $\alpha \sim \beta$, and even if $\alpha \wedge \alpha' \not\sim \beta$ in the nonmonotonic sense of (2), we prefer those models where $\alpha \wedge \alpha' \sim \beta$, but exceptions are possible by nonmonotonicity itself, as, e.g., $\alpha \wedge \alpha' \sim \neg\beta$ in (2).

   We minimize those exceptions, and resolve conflicts whenever possible, as in Fact 5.2.9 (page 88), by using the same principle as in level (2): we keep exception sets small. This is summarized in the specificity criterion.

(We might add a modified length of path criterion as follows: Let $x_0 \to x_1 \to \ldots \to x_n$, $x_i \to y_i$, $x_{i+1} \not\to y_i$. We know by Fact 5.2.9 (page 88) that $x_0 < \ldots < x_n$, then any shorter chain $a \to \ldots \to b$ has a shorter possible size reduction (if there are no other chains, of course!), and we can work with this. This is the same concept as in [Sch18e], section 4.)

This has again a clear (preferential) semantics, as our characterisations are abstract, see e.g. [Sch18a].

Inductive reasoning is the upward version, the problem is to find the cases where it holds.

Analogical reasoning is the upward version going from X to $X \cup X'$.

The author recently discovered (reading [SEP13], section 4.3) that J. M. Keynes's Principle of the Limitation of Independent Variety, see [Key21] expresses essentially the same idea as homogenousness. (It seems, however, that the epistemological aspect, the naturalness of our concepts, is missing in his work.) By the way, [SEP13] also mentions "inference pressure" (in section 3.5.1) discussed in [Sch97-2], section 1.3.4, page 10. Thus, the ideas are quite interwoven.

"Uniformity of nature" is a concept similar to our homogenousness, see [Wik18c], and [Sim63]. The former is, however, usually applied to changes in time and place, and not to subsets, as we do.

5.1.2   Contributions of the Present Text

(1) Our main formal contribution here is to analyse various size relations between sets, see Section 5.2 (page 80), in particular when this relation is generated itself by a relation $\prec$ between elements, see Section 5.2.3 (page 95), - similarly to Definition 2.6 and Fact 2.7 in [Sch97-2].

These relations can then be used as semi-quantitative distances between sets, to calculate the "seriousness" of homogeneity violations.

Ideas and proofs are elementary.

(2) Conceptually, we consider

(2.1) Background logic, often non-monotonic, in some fixed language

(2.2) Reference classes, usually a subset of the formulas in the fixed language. They may be closed under operations like $\cap$, but not necessarily

(2.3) Notion of consistency of the background logic

(2.4) Specificity to solve conflicts in the background logic between properties of different reference classes, often based on a notion of distance

(2.5) We do not treat things like: related classes, properties, this is left to "knowledge engineering".

It is important to carefully select the predicates treated.

For instance, if X is the set of all vertebrate, Y the set of all mammals, Y' the set of all cats, then it is plausible that elements of Y' have many more additional properties to those valid in X than elements of Y do. Taxonomies are made exactly for this purpose, and good natural categories behave this way.

Compare the author's favorite example of bad categories:

Example 5.1.1

Enumerate the objects of the universe, and consider the class of those objects, whose number ends by 3. Now consider the subclasses of all objects whose number ends by 33, by 333, etc. We will not expect these concepts to have reasonable properties - apart from trivial ones.

The separation of reference classes and properties existed e.g. in KL-ONE.

Definition 5.1.2

(1) A similarity structure consists of a set of models or set of sets of models (for some fixed language), called possible reference classes - we do not assume any closure conditions -, and a model or set of models (not in the set of possible reference classes) T, called the target class, a set of formulas in the fixed language, called the possible properties, and a partial order $\leq$ on the reference classes relative to the target class. (Intuitively, this partial order chooses the "best" reference classes wrt. the target class.) For simplicity, assume that there are no infinite descending chains in the order.

(2) A similarity theory consists of a similarity structure, together with a procedure to solve contradictions. More precisely, if two or more reference classes, all $\leq$-optimal, contradict each other about ϕ, we need to know what to do.

Several possibilities come to mind (there might be still others):

(2.1) branch into different "extensions" of T, with, e.g. $T \models \phi$ and $T \models \neg\phi$,

(2.2) give no information about T and ϕ (direct scepticism)

(2.3) if ϕ permits more than two truth values, like numerical values, find a compromise (which may depend on the number of votes for the different values).

This is also possible for non-monotonic logics, where (instead of above models, we have now sets of models for reference and target classes) $R \models \nabla\phi$ (meaning: almost everywhere in R, ϕ holds) and $R' \models \nabla\neg\phi$, we conclude for the target class $T \models \heartsuit\phi$, meaning: the subset of T where ϕ holds, has medium size.

(3) A homogenousness structure is a similarity structure (the set version), with $\subseteq$ as basic relation ($\subseteq$ may be strict, or "soft", with exceptions, with some underlying theory of exceptions). Possible reference classes have to be $\subseteq$-above the target class, i.e. $T \subseteq R$ for each R. The partial order is again $\subseteq$, perhaps with some embellishments to solve more conflicts, see Chapter 11, Formal Construction, in [GS16]. (E.g., if we cannot compare by $\subseteq$, we may resort to some variant of path length.)

The more specific superclass will win, e.g., if $T \subseteq R \subseteq R'$, then R will win over R'.

5.2 Formal Properties

5.2.1 Basic Definitions and Facts

Remark 5.2.1

Note that we use in the proof of Fact 5.2.5 (page 83) a - to the author - new way of writing down proofs which makes them almost mechanical, comparable to elementary school maths, eliminating terms on both sides of an expression. The idea is to write down all sets involved, this is tedious, but the rest is trivial.

Definition 5.2.1

Let $X \neq \emptyset$.

(1) $\mathcal{F}(X) \subseteq \mathcal{P}(X)$ is called a filter on X iff

(F1) $X \in \mathcal{F}(X)$, $\emptyset \notin \mathcal{F}(X)$

(F2) $A \subseteq B \subseteq X$, $A \in \mathcal{F}(X) \Rightarrow B \in \mathcal{F}(X)$

(F3) $A, B \in \mathcal{F}(X) \Rightarrow A \cap B \in \mathcal{F}(X)$ (finite intersection suffices here)

(2) If there is $A \subseteq X$ such that $\mathcal{F}(X) = \{A' \subseteq X : A \subseteq A'\}$, we say that $\mathcal{F}(X)$ is the (principal) filter generated by A. For historical reasons, we will often note this A $\mu(X)$.

(3) $\mathcal{I}(X) \subseteq \mathcal{P}(X)$ is called an ideal on X iff

(I1) $X \notin \mathcal{I}(X)$, $\emptyset \in \mathcal{I}(X)$

(I2) $A \subseteq B \subseteq X$, $B \in \mathcal{I}(X) \Rightarrow A \in \mathcal{I}(X)$

(I3) $A, B \in \mathcal{I}(X) \Rightarrow A \cup B \in \mathcal{I}(X)$ (finite union suffices here)

Intuitively, filters over X contain big subsets of X, ideals small subsets.

Thus, (F1), (F2), (I1), (I2) are natural properties, (F3) and (I3) give algebraic strength. We sometimes emphasize the number of times the latter (and other properties to be discussed below) were used in a proof.

Definition 5.2.2

Let $X \neq \emptyset$.

If $\mathcal{F}(X)$ is a filter over X, then

$\{A \subseteq X : X - A \in \mathcal{F}(X)\}$ is the corresponding ideal $\mathcal{I}(X)$,

and, conversely

if $\mathcal{I}(X)$ is an ideal over X, then

$\{A \subseteq X : X - A \in \mathcal{I}(X)\}$ is the corresponding filter $\mathcal{F}(X)$.

When we go from filter to ideal to filter over fixed X, we always mean the corresponding structure.

Given $\mathcal{F}(X)$ and the corresponding $\mathcal{I}(X)$, we set

$\mathcal{M}(X) := \{A \subseteq X : A \notin \mathcal{F}(X) \cup \mathcal{I}(X)\}$ - the set of medium size subsets.

We now define several coherence properties - properties relating filters and ideals over different base sets X, Y, etc.

Definition 5.2.3

(Coh1) $X \subseteq Y \Rightarrow \mathcal{I}(X) \subseteq \mathcal{I}(Y)$.

(Coh2) $X \in \mathcal{F}(Y)$, $A \in \mathcal{I}(Y) \Rightarrow A \cap X \in \mathcal{I}(X)$

(Coh2a) $Z, Z' \in \mathcal{I}(B) \Rightarrow Z - Z' \in \mathcal{I}(B - Z')$

(Coh-RK) $X \subseteq Y, X \notin \mathcal{I}(Y), A \in \mathcal{I}(Y) \Rightarrow A \cap X \in \mathcal{I}(X)$

(Coh1) is, by the intuition of an ideal, a very natural property, and will not be mentioned in proofs.

For principal filters, we define:

(μPR) $X \subseteq Y \Rightarrow \mu(Y) \cap X \subseteq \mu(X)$

(μCUM) $\mu(Y) \subseteq X \subseteq Y \Rightarrow \mu(X) = \mu(Y)$

(μRK) $X \subseteq Y, \mu(Y) \cap X \neq \emptyset \Rightarrow \mu(Y) \cap X = \mu(X)$

Working with principal filters gives us easy examples, drawing simple diagrams is often sufficient as an (idea for) a proof.

Fact 5.2.2

(Coh2) and (Coh2a) are equivalent.

Proof

"$(Coh2a) \Rightarrow (Coh2)$":

Let $X \in \mathcal{F}(Y), A \in \mathcal{I}(Y)$, then $Y - X \in \mathcal{I}(Y)$, so $A \cap X = A - (Y - X) \in \mathcal{I}(Y - (Y - X)) = \mathcal{I}(X)$

"$(Coh2) \Rightarrow (Coh2a)$":

$Z, Z' \in \mathcal{I}(B) \Rightarrow B - Z' \in \mathcal{F}(B) \Rightarrow Z - Z' = Z \cap (B - Z') \in \mathcal{I}(B - Z')$

$\square$

We now show further coherence properties in Fact 5.2.3 (page 82) through Fact 5.2.9 (page 88).

Fact 5.2.3

(1) $A, A' \subseteq B, A \Delta A' \in \mathcal{I}(B), A \in \mathcal{F}(B) \rightarrow A' \in \mathcal{F}(B)$.

(2) $A, A' \subseteq B, A \Delta A' \in \mathcal{I}(B), A \in \mathcal{I}(B) \rightarrow A' \in \mathcal{I}(B)$.

Proof

(1) $A \Delta A' \in \mathcal{I}(B) \rightarrow (B - (A \cup A')) \cup (A \cap A') = B - (A \Delta A') \in \mathcal{F}(B)$. By $A \in \mathcal{F}(B)$, $A \cap ((B - (A \cup A')) \cup (A \cap A')) = A \cap A' \in \mathcal{F}(B)$, so $A' \in \mathcal{F}(B)$.

(2) Note that $A \Delta A' = (B - A) \Delta (B - A')$. Thus consider B-A, $B - A'$, and apply (1).

Fact 5.2.4

Let (Coh1) and (Coh2a) hold.

Let $X \subseteq Y$, $X \Delta X' \in \mathcal{I}(X \cup X')$, then:

(1) $X \in \mathcal{I}(Y)$ iff $X' \in \mathcal{I}(Y \cup X')$

(2) $X \in \mathcal{F}(Y)$ iff $X' \in \mathcal{F}(Y \cup X')$

(3) $X \in \mathcal{M}(Y)$ iff $X' \in \mathcal{M}(Y \cup X')$

Proof

By $X \subseteq Y$ and (Coh1) $X \Delta X' \in \mathcal{I}(Y \cup X')$.

(1)

"$\Rightarrow$": $X \in \mathcal{I}(Y) \subseteq \mathcal{I}(Y \cup X')$, $(X \Delta X') \in \mathcal{I}(X \cup X') \subseteq \mathcal{I}(Y \cup X')$, so $X' \in \mathcal{I}(Y \cup X')$ by Fact 5.2.3 (page 82).

"$\Leftarrow$": $X' - Y \subseteq X' \in \mathcal{I}(Y \cup X')$, so by (Coh2a) $X = X - (X' - Y) \in \mathcal{I}((Y \cup X') - (X' - Y)) = \mathcal{I}(Y)$.

(2)

"$\Leftarrow$": $X' \in \mathcal{F}(Y \cup X')$, $X \Delta X' \in \mathcal{I}(X \cup X') \subseteq \mathcal{I}(Y \cup X')$, so by Fact 5.2.3 (page 82) $X \in \mathcal{F}(Y \cup X')$, so $X \in \mathcal{F}(Y)$.

"$\Rightarrow$": $X \in \mathcal{F}(Y) \Rightarrow Y - X \in \mathcal{I}(Y) \subseteq \mathcal{I}(Y \cup X')$. $(Y - X) \Delta (Y - X') = X \Delta X' \in \mathcal{I}(X \cup X') \subseteq \mathcal{I}(Y \cup X')$. Thus, $Y - X' \in \mathcal{I}(Y \cup X')$ by Fact 5.2.3 (page 82), so $X' = (Y \cup X') - (Y - X') \in \mathcal{F}(Y \cup X')$.

(3)

Suppose $X \in \mathcal{M}(Y)$, but $X' \in \mathcal{I}(Y \cup X')$, then $X \in \mathcal{I}(Y)$, contradiction. The other cases are analogous.

□

The strategy is essentially the same for the following Facts. We go up to $Y \cup X \cup X'$, and again down to $Y \cup X$ of $Y \cup X'$, using $X \Delta X' \in \mathcal{I}(X \cup X' \cup Y)$, and thus $X \in \mathcal{I}(X \cup X' \cup Y)$ iff $x' \in \mathcal{I}(X \cup X' \cup Y)$.

Roughly, the argument is that changing things a little has no influence, see Fact 5.2.3 (page 82).

We give now alternative proofs, which are rather mechnical, but follow the same strategy.

Fact 5.2.5

Let $X - Y \in \mathcal{I}(X)$, $X \Delta X' \in \mathcal{I}(X \cup X')$, then:

(1) $X \in \mathcal{I}(Y \cup X)$ iff $X' \in \mathcal{I}(Y \cup X')$

(2) $X \in \mathcal{F}(Y \cup X)$ iff $X' \in \mathcal{F}(Y \cup X')$

(3) $X \in \mathcal{M}(Y \cup X)$ iff $X' \in \mathcal{M}(Y \cup X')$

Proof

We simplify notation, using + and - for set union and difference, e.g. $-X + X' - Y$ will mean $(X' - X)$-Y.

Define

$A := +X + X' + Y$

$B := +X + X' - Y$

$C := +X - X' + Y$

$D := +X - X' - Y$

$E := -X + X' + Y$

$F := -X + X' - Y$

$G := -X - X' + Y$

$H := -X - X' - Y$

Let

$X := ABCD$ (for $A + B + C + D$ etc.)

$X' := ABEF$

$Y := ACEG$.

Then

$X - Y = BD$

$X \cup X' = ABCDEF$

$X \Delta X' = CDEF$

$Y \cup X = ABCDEG$

$Y \cup X' = ABCEFG$

$(Y \cup X) - X = EG$

$(Y \cup X') - X' = CG$

The prerequisites are:

(1) $BD \in \mathcal{I}(ABCD)$

(2) $CDEF \in \mathcal{I}(ABCDEF)$

(1) We have to show $X \in \mathcal{I}(Y \cup X)$ iff $X' \in \mathcal{I}(Y \cup X')$, i.e.

(3) $ABCD \in \mathcal{I}(ABCDEG)$

iff

(4) $ABEF \in \mathcal{I}(ABCEFG)$

(3) $\Rightarrow$ (4):

By (3), ABCD $\in \mathcal{I}(ABCDEG) \subseteq \mathcal{I}(ABCDEFG)$, by (2), CDEF $\in \mathcal{I}(ABCDEF) \subseteq \mathcal{I}(ABCDEFG)$, so ABCDEF $\in \mathcal{I}(ABCDEFG)$, so ABEF $\subseteq$ ABCEF $\in \mathcal{I}(ABCEFG)$.

(4) $\Rightarrow$ (3):

Analogously:

By (4), ABEF $\in \mathcal{I}(ABCEFG) \subseteq \mathcal{I}(ABCDEFG)$, by (2), CDEF $\in$ $\mathcal{I}(ABCDEF) \subseteq \mathcal{I}(ABCDEFG)$, so ABCDEF $\in \mathcal{I}(ABCDEFG)$, so ABCD $\subseteq$ ABCDE $\in \mathcal{I}(ABCDEG)$.

(2) (Similarly) $X \in \mathcal{F}(X \cup Y)$ iff $X' \in \mathcal{F}(Y \cup X')$, or $(X \cup Y) - X \in \mathcal{I}(X \cup Y)$ iff $(Y \cup X') - X' \in \mathcal{I}(Y \cup X')$, i.e.

(3) EG $\in \mathcal{I}(ABCDEG)$ iff (4) CG $\in \mathcal{I}(ABCEFG)$.

(3) $\Rightarrow$ (4) :

By (3), EG $\in \mathcal{I}(ABCDEFG)$, by (2) CDEF $\in \mathcal{I}(ABCDEFG)$, so CDEFG $\in \mathcal{I}(ABCDEFG)$, and CEFG $\in \mathcal{I}(ABCEFG)$, and CG $\in \mathcal{I}(ABCEFG)$.

(4) $\Rightarrow$ (3) :

By (4), CG $\in \mathcal{I}(ABCDEFG)$, by (2) CDEF $\in \mathcal{I}(ABCDEFG)$, so CDEFG $\in \mathcal{I}(ABCDEFG)$, and CDEG $\in \mathcal{I}(ABCDEG)$, and EG $\in \mathcal{I}(ABCDEG)$.

(3) As above: $X \in \mathcal{M}(Y \cup X)$ iff $X' \in \mathcal{M}(Y \cup X')$.

Suppose e.g. $X \in \mathcal{M}(Y \cup X)$ and $X' \in \mathcal{I}(Y \cup X')$, then $X \in \mathcal{I}(Y \cup X)$, contradiction.

$\square$

Remark 5.2.6

Consider the proof of (1) above. The important properties are: the right hand side of properties (3) and (4) differ in D (in (3)) and F (*in*(4)). Property (2) has D and F on both sides, CDEF $\in \mathcal{I}(ABCDEF)$, which allows (after taking unions) to eliminate both D and F.

Similar properties hold below in Fact 5.2.7 (page 85).

Fact 5.2.7

Let (1) $X - Y \in \mathcal{I}(X)$, (2) $Y \triangle Y' \in \mathcal{I}(Y \cup Y')$, then:

(1) $X \in \mathcal{I}(Y \cup X)$ iff $X \in \mathcal{I}(Y' \cup X)$

(2) $X \in \mathcal{F}(Y \cup X)$ iff $X \in \mathcal{F}(Y' \cup X)$

(3) $X \in \mathcal{M}(Y \cup X)$ iff $X \in \mathcal{M}(Y' \cup X)$

Proof

$A := +X + Y + Y'$

$B := +X + Y - Y'$

$C := +X - Y + Y'$

$D := +X - Y - Y'$

$E := -X + Y + Y'$

$F := -X + Y - Y'$

$G := -X - Y + Y'$

$H := -X - Y - Y'$

Let

$X := ABCD$

$Y := ABEF$

$Y' := ACEG.$

Then

$X - Y = CD$

$X - Y' = BD$

$Y \Delta Y' = BCFG$

$Y \cup Y' = ABCEFG$

$Y \cup X = ABCDEF$

$Y' \cup X = ABCDEG$

$(Y \cup X) - X = \text{EF}$

$(Y' \cup X) - X = \text{EG}$

The prerequisites are:

(1) $CD \in \mathcal{I}(ABCD)$

(2) $BCFG \in \mathcal{I}(ABCEFG)$

(1) We have to show $X \in \mathcal{I}(Y \cup X)$ iff $X \in \mathcal{I}(Y' \cup X)$ i.e., that (3) and (4) are equivalent.

(3) $ABCD \in \mathcal{I}(ABCDEF)$

(4) $ABCD \in \mathcal{I}(ABCDEG)$

(3) $\Rightarrow$ (4):

$BCFG \in \mathcal{I}(ABCEFG) \subseteq \mathcal{I}(ABCDEFG)$

$ABCD \in \mathcal{I}(ABCDEF) \subseteq \mathcal{I}(ABCDEFG)$

so

$ABCDFG \in \mathcal{I}(ABCDEFG)$, so $ABCD \in \mathcal{I}(ABCDE) \subseteq \mathcal{I}(ABCDEG)$

(4) $\Rightarrow$ (3):

BCFG $\in \mathcal{I}(ABCEFG) \subseteq \mathcal{I}(ABCDEFG)$

ABCD $\in \mathcal{I}(ABCDEG) \subseteq \mathcal{I}(ABCDEFG)$, so

ABCDFG $\in \mathcal{I}(ABCDEFG)$, so ABCD $\in \mathcal{I}(ABCDE) \subseteq \mathcal{I}(ABCDEF)$

(2) We have to show $X \in \mathcal{F}(Y \cup X)$ iff $X \in \mathcal{F}(Y' \cup X)$, i.e. that $(Y \cup X) - X \in \mathcal{I}(Y \cup X)$ iff $(Y' \cup X) - X \in \mathcal{I}(Y' \cup X)$, i.e.,

(3) EF $\in \mathcal{I}(ABCDEF)$ iff (4) EG $\in \mathcal{I}(ABCDEG)$.

(3) $\Rightarrow$ (4) :

By (3) EF $\in \mathcal{I}(ABCDEFG)$, by (2) BCFG $\in \mathcal{I}(ABCDEFG)$, so BCEFG $\in \mathcal{I}(ABCDEFG)$, so EG $\subseteq$ BCEG $\in \mathcal{I}(ABCDEG)$.

(4) $\Rightarrow$ (3) : By (4) EG $\in \mathcal{I}(ABCDEFG)$, by (2) BCFG $\in \mathcal{I}(ABCDEFG)$, so BCEFG $\in \mathcal{I}(ABCDEFG)$, so EF $\subseteq$ BCEF $\in \mathcal{I}(ABCDEF)$.

(3) $X \in \mathcal{M}(Y \cup X)$ iff $X \in \mathcal{M}(Y' \cup X)$:

As the corrsponding result for $\mathcal{M}$ in Fact 5.2.5 (page 83).

$\square$

Fact 5.2.8

If $X \in \mathcal{F}(X')$, then $(X \cap A \in \mathcal{F}(X) \Leftrightarrow X' \cap A \in \mathcal{F}(X'))$

Proof

"$\Rightarrow$": $X' - X \in \mathcal{I}(X')$, $X - A \in \mathcal{I}(X) \subseteq \mathcal{I}(X')$, so $(X' - X) \cup (X - A) \in \mathcal{I}(X')$
$\Rightarrow X' - ((X' - X) \cup (X - A)) = A \in \mathcal{F}(X')$.

"$\Leftarrow$": $X' \cap A \in \mathcal{F}(X')$, $X' - X \in \mathcal{I}(X') \Rightarrow X \cap A = (X' \cap A) \cap X \in \mathcal{F}(X)$.

Second proof:

Set $B := (X' - X)$-A, $C := (X' - X) \cap A$, $D := X$-A, $E := X \cap A$.

We want to show: if $D \cup E \in \mathcal{F}(B \cup C \cup D \cup E)$, then $(E \in \mathcal{F}(E \cup D) \Leftrightarrow E \cup C \in \mathcal{F}(B \cup C \cup D \cup E))$, equivalently if $B \cup C \in \mathcal{I}(B \cup C \cup D \cup E)$, then $D \in \mathcal{I}(E \cup D) \Leftrightarrow B \cup D \in \mathcal{I}(B \cup C \cup D \cup E))$

Let $B \cup D \in \mathcal{I}(B \cup C \cup D \cup E)$, by $C \in \mathcal{I}(B \cup C \cup D \cup E)$, $B \cup C \cup D \in \mathcal{I}(B \cup C \cup D \cup E)$, so $D \in \mathcal{I}(D \cup E)$.

Conversely, let $D \in \mathcal{I}(D \cup E) \subseteq \mathcal{I}(B \cup C \cup D \cup E)$, by $B \cup C \in \mathcal{I}(B \cup C \cup D \cup E)$, $B \cup D \in \mathcal{I}(B \cup C \cup D \cup E)$.

$\square$

Fact 5.2.9

Let $(X - Y) \in \mathcal{I}(X)$, $(X \cap Z) \in \mathcal{I}(X)$, $(Y - Z) \in \mathcal{I}(Y)$, then $X \in \mathcal{I}(X \cup Y)$.

Thus, if $\neg\phi$ holds mostly in X, ϕ holds mostly in Y, most of X is in Y, then $X < Y$ (see Definition 5.2.5 (page 90)), and in inheritance notation, if $X \to Y$, $X \not\to Z$, $Y \to Z$, then $X < Y$, and $Y \not\to X$.

Proof

$A := +X + Y + Z$

$B := +X + Y - Z$

$C := +X - Y + Z$

$D := +X - Y - Z$

$E := -X + Y + Z$

$F := -X + Y - Z$

$G := -X - Y + Z$

$H := -X - Y - Z$

Let

$X := ABCD$

$Y := ABEF$

$Z := ACEG.$

Then

$X - Y = CD$

$X \cap Z = AC$

$Y - Z = BF$

$X \cup Y = ABCDEF.$

The prerequisites are:

$CD \in \mathcal{I}(ABCD) \subseteq \mathcal{I}(ABCDEF)$

$AC \in \mathcal{I}(ABCD) \subseteq \mathcal{I}(ABCDEF)$

$BF \in \mathcal{I}(ABEF) \subseteq \mathcal{I}(ABCDEF)$, so

$ABCD \subseteq ABCDF \in \mathcal{I}(ABCDEF).$

$\square$

5.2.1.1 Remarks on Principal and Relation Generated Filters

Fact 5.2.10

Let $\mathcal{F}(A) := \{A' \subseteq A : \mu(A) \subseteq A'\}$ the principal filter over A generated by $\mu(A)$, then the corresponding $\mathcal{I}(A) = \{A' \subseteq A : A' \cap \mu(A) = \emptyset\}$, and $\mathcal{M}(A) = \{A' \subseteq A : A' \cap \mu(A) \neq \emptyset,$ and $\mu(A) \not\subseteq A'\}$. $\square$

Fact 5.2.11

Let the filters be principal filters.

(1) (Coh1) is equivalent to (μPR).

(2) (μCum) implies (Coh2), and $(Coh1) + (Coh2)$ imply (μCum).

(3) (μRK) implies (Coh-RK), and $(Coh1) + (Coh\text{-}RK)$ imply (μRK).

Proof

(1) $(\mu PR) \Rightarrow$ (Coh1): $A \in \mathcal{I}(X) \to A \cap \mu(X) = \emptyset \to A \cap \mu(Y) = \emptyset \to A \in \mathcal{I}(Y)$.

(Coh1) $\Rightarrow (\mu PR)$: Suppose there is $X \subseteq Y$ such that (μPR) fails, so $\mu(Y) \cap X \not\subseteq \mu(X)$, then $X - \mu(X) \in \mathcal{I}(X)$, but $(X - \mu(X)) \cap \mu(Y) \neq \emptyset$, so $X - \mu(X) \notin \mathcal{I}(Y)$.

(2) $(\mu CUM) \Rightarrow$ (Coh2): Let $A, B \in \mathcal{I}(X)$, $A \cap B = \emptyset$, so $\mu(X - B) = \mu(X) \Rightarrow A \in \mathcal{I}(X - B)$.

$(Coh1) + (Coh2) \Rightarrow (\mu CUM)$: Let $\mu(X) \subseteq Y \subseteq X$. $X - Y$, $Y - \mu(X) \in \mathcal{I}(X)$, and $(X - Y) \cap (Y - \mu(X)) = \emptyset$, so $Y - \mu(X) \in \mathcal{I}(X - (X - Y)) = \mathcal{I}(Y)$, so $\mu(Y) \subseteq \mu(X)$. $\mu(X) \subseteq \mu(Y)$ follows from (Coh1)

(3) $(\mu RK) \Rightarrow$ (Coh-RK): Let $A \in \mathcal{I}(Y) \to A \cap \mu(Y) = \emptyset \to A \cap \mu(X) = \emptyset \to A \in \mathcal{I}(X)$.

$(Coh1) + (Coh\text{-}RK) \Rightarrow (\mu RK)$: Let $X \subseteq Y$ and $X \notin \mathcal{I}(Y)$, $(Y - \mu(Y)) \in \mathcal{I}(Y)$, so by (Coh-RK) $(Y - \mu(Y)) \cap X \in \mathcal{I}(X)$, so $(X - (Y - \mu(Y))) \in \mathcal{F}(X)$ and $\mu(X) \subseteq (X - (Y - \mu(Y))) = \mu(Y) \cap X$, but by (Coh1) $\mu(Y) \cap X \subseteq \mu(X)$.

$\square$

We now consider filters generated by preferential structures.

Definition 5.2.4

Let $X \neq \emptyset$, $\prec$ a binary relation on X, we define for $\emptyset \neq A \subseteq X$

$\mu(A) := \{x \in A : \neg \exists x' \in A.x' \prec x\}$

(This is simplified definition, without "copies", see e.g. [Sch18] for the full picture.)

We assume in the sequel that for any such X and A, $\mu(A) \neq \emptyset$.

We define the following standard properties for the relation $\prec$:

(1) Transitivity (trivial)

(2) Smoothness

If $x \in X - \mu(X)$, there there is $x' \in \mu(X).x' \prec x$

(3) Rankedness

If neither $x \prec x'$ nor $x' \prec x$, and $x \prec y$ $(y \prec x)$, then also $x' \prec y$ $(y \prec x')$.

(Rankedness implies transitivity.)

See, e.g. Chapter 1 in [Sch18a].

Remark 5.2.12

(Simplified)

(μPR) characterizes general preferential structures,

(μCUM) characterizes smooth preferential structures,

(μRK) characterizes ranked preferential structures.

See [Sch18] for details.

5.2.2   More Detailed Size Comparisons

We consider here more detailed size comparisons, and comparisons relative to some given set X.

Definition 5.2.5

Given X, $\mathcal{F}(X)$ (and corresponding $\mathcal{I}(X)$, $\mathcal{M}(X)$), and $A, B \subseteq X$, we define:

(1) $X \sqsubset A$ iff X is a small subset of A, i.e. $X \in \mathcal{I}(A)$,

(2) $X \sqsubset_n A$ iff there are $A_i, 1 \le i \le n-1$ such that $X \sqsubset A_1 \sqsubset A_2 \sqsubset \ldots \sqsubset A_{n-1} \sqsubset A$,

   The index n says how much smaller X is compared to A.

(3) $A <_X B :\Leftrightarrow A \in \mathcal{I}(X), B \in \mathcal{F}(X)$

(4) $A <'_X B :\Leftrightarrow$
   (a) $B \in \mathcal{F}(X)$ and $A \in \mathcal{I}(X) \cup \mathcal{M}(X)$
   or
   (b) $B \in \mathcal{M}(X)$ and $A \in \mathcal{I}(X)$

(5) If $X = A \cup B$, we write $A < B$ and $A <' B$, instead of $A <_X B$ and $A <'_X B$.
   (Note: $X \in \mathcal{I}(X \cup Y) \Leftrightarrow X \in \mathcal{I}(X \cup Y) \wedge Y \in \mathcal{F}(X \cup Y)$)

(6) $X <_n A$ iff there are $A_i, 1 \le i \le n-1$ such that $X < A_1 < A_2 < \ldots < A_{n-1} < A$.

Remark 5.2.13

$X \in \mathcal{I}(X \cup Y) \Rightarrow Y \in \mathcal{F}(X \cup Y)$, but not necessarily the converse.

Proof

$X \in \mathcal{I}(X \cup Y) \Rightarrow (X \cup Y) - X \in \mathcal{F}(X \cup Y)$, and $(X \cup Y) - X \subseteq Y$, so $Y \in \mathcal{F}(X \cup Y)$.

For the converse: Consider $X = Y$, then $Y \in \mathcal{F}(X \cup Y)$, but $X \notin \mathcal{I}(X \cup Y)$.

□

Note that case $(4)(b)$ of the definition is impossible if $X = A \cup B$: By Remark 5.2.13 (page 90), if $A \in \mathcal{I}(A \cup B)$, then $B \in \mathcal{F}(A \cup B)$.

We will sometimes count applications of (I3) and (Coh2) - equivalently (Coh2a) -, see Definition 5.2.1 (page 81) and Definition 5.2.3 (page 81), using the notation (ns), for n applications of smaller ("s" for "size" or "smaller").

Fact 5.2.14

$<$ is transitive

Proof

Let $X < Y < Z$, so $X \in \mathcal{I}(X \cup Y)$ and $Y \in \mathcal{I}(Y \cup Z)$. We have to show $X \in \mathcal{I}(X \cup Z)$.

Consider $X \cup Y \cup Z$, then by $X \in \mathcal{I}(X \cup Y)$, $X \in \mathcal{I}(X \cup Y \cup Z)$. By the same argument, $Y \in \mathcal{I}(X \cup Y \cup Z)$, thus $Y - (X \cup Z) \in \mathcal{I}(X \cup Y \cup Z)$. As $(X \cup Y \cup Z) - (Y - (X \cup Z)) = X \cup Z$, and $X \cap (Y - (X \cup Z)) = \emptyset$, $X \in \mathcal{I}(X \cup Z)$ by Definition 5.2.3 (page 81), (Coh2a). □

Fact 5.2.15

(See [Sch18].)

(1) $A \sqsubset B \sqsubset C \Rightarrow A \sqsubset C$ (without any (ns))

(2) $A \sqsubset B \sqsubset A$ is impossible

(3) $A \subseteq B < C \Rightarrow A < C$ (1s)

(4) $A < B \subseteq C \Rightarrow A < C$ (without any (ns))

(5) $X < Y < Z \Rightarrow X < Z$ (1s)

(6) $X < Y < X$ is impossible (1s)

(7) $X < Y < Z < U \Rightarrow X < U$ (2s)

(8) $X_0 < \ldots < X_n \Rightarrow X_0 < X_n$ by $((n-1) * s)$

(9) $X < Y \Rightarrow X \cap Y \sqsubset Y$ (1s)

(10) $X \cap Y \sqsubset X \cup Y$, $X - Y \sqsubset X \Rightarrow$ (1s) $X \cap Y \sqsubset Y$

Proof

(1) By (I2) or (Coh1).

(2) By (1).

(3) By $B \sqsubset B \cup C$ and $B - A \subseteq B$, we have $B - A \sqsubset B \cup C$ by $(I \subseteq)$. So by (Coh2a) $A = B - (B - A) \sqsubset B \cup C - (B - A) \subseteq A \cup C$, the latter by (Coh1).

(4) $A \sqsubset A \cup B \subseteq A \cup C \Rightarrow A \sqsubset A \cup C$ by (Coh1).

(5) We have $X \sqsubset X \cup Y$, $Y \sqsubset Y \cup Z$, so $X \sqsubset X \cup Y \cup Z$ and $Y \sqsubset X \cup Y \cup Z$, so $Y - (X \cup Z) \sqsubset X \cup Y \cup Z$. So (1s) $X = X - (Y - (X \cup Z)) \sqsubset X \cup Y \cup Z - (Y - (X \cup Z)) = X \cup Z$.

(6) By (6) and (7).

(7) By $X \sqsubset X \cup Y$, $Y \sqsubset Y \cup Z$, $Z \sqsubset X \cup U$, so $Y \cup Z \sqsubset X \cup Y \cup Z \cup U$ (1s). So $Y \cup Z - X \cup U \subseteq Y \cup Z \sqsubset X \cup Y \cup Z \cup U$. $X \cup U = X \cup Y \cup Z \cup U - (Y \cup Z - X \cup U)$, $X = X - (Y \cup Z - X \cup U)$. So $X \sqsubset X \cup U$ (1s).

(8) Analogous to (7).

(9) $X < Y \Rightarrow X \sqsubset X \cup Y$, so $X - Y \sqsubset X \cup Y = Y \cup (X - Y)$, and $X = (X \cap Y) \cup (X - Y)$. Thus (1s) $X \cap Y \sqsubset Y$.

(10) $X - Y \sqsubset X \subseteq X \cup Y \Rightarrow$ (1s) $X \cap Y = X \cap Y - (X - Y) \sqsubset X \cup Y - (X - Y) = Y$.

$\square$

We mention the following without proofs, it is very close to Fact 5.2.5 (page 83).

Fact 5.2.16

$(Coh1) + (Coh2)$ imply:

(1) Let $X \in \mathcal{F}(X')$, then $(X \cap A \in \mathcal{F}(X) \Leftrightarrow X' \cap A \in \mathcal{F}(X'))$

(2) Let $X' \in \mathcal{F}(X)$, $Y' \in \mathcal{F}(Y)$, then the following four conditions are equivalent: $X < Y$, $X' < Y$, $X < Y'$, $X' < Y'$

Fact 5.2.17

If (Coh-RK) holds, then $<'$ is transitive.

Proof

First, if $A \in \mathcal{I}(A \cup B)$, then $B \in \mathcal{F}(A \cup B)$, so we need not consider case $(4)(b)$ in Definition 5.2.5 (page 90).

Second, we will use repeatedly:

$(*)$ If $RSTU \in \mathcal{M}(RSTUVW)$, and $RSVW \in \mathcal{F}(RSTUVW)$, so $TU \in \mathcal{I}(RSTUVW)$, then by (Coh-RK) $TU \in \mathcal{I}(RSTU)$, and thus also $T \in \mathcal{I}(RST)$ and $U \in \mathcal{I}(RSU)$ by (Coh2).

Consider

$A := +X + Y + Z$

$B := +X + Y - Z$

$C := +X - Y + Z$

$D := +X - Y - Z$

$E := -X + Y + Z$

$F := -X + Y - Z$

$G := -X - Y + Z$

$H := -X - Y - Z$

and

$X := ABCD$

$Y := ABEF$

$Z := ACEG.$

Then

$X \cup Y = ABCDEF$

$X \cup Z = ABCDEG$

$Y \cup Z = ABCEFG.$

There are 4 cases to consider:

(a)
$$X \in \mathcal{I}(X \cup Y), \, Y \in \mathcal{F}(X \cup Y), \, Y \in \mathcal{I}(Y \cup Z), \, Z \in \mathcal{F}(Y \cup Z)$$

(b)
$$X \in \mathcal{M}(X \cup Y), \, Y \in \mathcal{F}(X \cup Y), \, Y \in \mathcal{M}(Y \cup Z), \, Z \in \mathcal{F}(Y \cup Z)$$

(c)
$$X \in \mathcal{M}(X \cup Y), \, Y \in \mathcal{F}(X \cup Y), \, Y \in \mathcal{I}(Y \cup Z), \, Z \in \mathcal{F}(Y \cup Z)$$

(d)
$$X \in \mathcal{I}(X \cup Y), \, Y \in \mathcal{F}(X \cup Y), \, Y \in \mathcal{M}(Y \cup Z), \, Z \in \mathcal{F}(Y \cup Z)$$

In each case we show $Z \in \mathcal{F}(X \cup Z)$, $X \notin \mathcal{F}(X \cup Z)$, i.e.

$ACEG \in \mathcal{F}(ABCDEG)$ and $ABCD \notin \mathcal{F}(ABCDEG)$.

The proofs are elementary and tedious.

Case (a)

This was done already in Fact 5.2.15 (page 91), (5).

Case (b)

$X \in \mathcal{M}(X \cup Y)$, $Y \in \mathcal{F}(X \cup Y)$, $Y \in \mathcal{M}(Y \cup Z)$, $Z \in \mathcal{F}(Y \cup Z)$, thus:

$ABCD \in \mathcal{M}(ABCDEF)$, $ABEF \in \mathcal{F}(ABCDEF)$, $ABEF \in \mathcal{M}(ABCEFG)$, $ACEG \in \mathcal{F}(ABCEFG)$.

We show

$ACEG \in \mathcal{F}(ABCDEG)$, $ABCD \notin \mathcal{F}(ABCDEG)$.

By applying (∗) twice, we have $CD \in \mathcal{I}(ABCD)$ and $BF \in \mathcal{I}(ABEF)$, thus $B \in \mathcal{I}(ABE)$ too.

Thus $BCD \in \mathcal{I}(ABCDE) \subseteq \mathcal{I}(ABCDEG)$, so $AEG \in \mathcal{F}(ABCDEG)$, so $ACEG \in \mathcal{F}(ABCDEG)$.

Suppose $ABCD \in \mathcal{F}(ABCDEG)$, so $EG \in \mathcal{I}(ABCDEG)$, so by $BCD \in \mathcal{I}(ABCDEG)$, $BCDEG \in \mathcal{I}(ABCDEG)$, so $BCEG \in \mathcal{I}(ABCEG) \subseteq \mathcal{I}(ABCEFG)$, so $CG \in \mathcal{I}(ABCEFG)$, and $ABEF \in \mathcal{F}(ABCEFG)$, contradiction.

Case (c)

$X \in \mathcal{M}(X \cup Y)$, $Y \in \mathcal{F}(X \cup Y)$, $Y \in \mathcal{I}(Y \cup Z)$, $Z \in \mathcal{F}(Y \cup Z)$, thus:

$ABCD \in \mathcal{M}(ABCDEF)$, $ABEF \in \mathcal{F}(ABCDEF)$, $ABEF \in \mathcal{I}(ABCEFG)$, $ACEG \in \mathcal{F}(ABCEFG)$.

By applying (∗), we have $CD \in \mathcal{I}(ABCD)$, thus $C \in \mathcal{I}(ABC)$ and $D \in \mathcal{I}(ABD)$. Thus, by $ABEF \in \mathcal{I}(ABCEFG)$, $ABCEF \in \mathcal{I}(ABCEFG)$, and $ABCE \in \mathcal{I}(ABCEG)$. Thus, $B \in \mathcal{I}(BG)$, $D \in \mathcal{I}(ABD)$, and $BD \in \mathcal{I}(ABCDEG)$, so $ACEG \in \mathcal{F}(ABCDEG)$.

On the other hand, $ABCE \in \mathcal{I}(ABCEG) \subseteq \mathcal{I}(ABCDEG)$ and by $D \in \mathcal{I}(ABD)$, $ABCD \in \mathcal{I}(ABCDEG)$.

Case (d)

$X \in \mathcal{I}(X \cup Y)$, $Y \in \mathcal{F}(X \cup Y)$, $Y \in \mathcal{M}(Y \cup Z)$, $Z \in \mathcal{F}(Y \cup Z)$, thus:

$ABCD \in \mathcal{I}(ABCDEF)$, $ABEF \in \mathcal{F}(ABCDEF)$, $ABEF \in \mathcal{M}(ABCEFG)$, $ACEG \in \mathcal{F}(ABCEFG)$.

By applying (∗), we have $BF \in \mathcal{I}(ABEF)$, thus $B \in \mathcal{I}(ABE)$ and $F \in \mathcal{I}(AEF)$.

By $ABCD \in \mathcal{I}(ABCDEF)$ and $F \in \mathcal{I}(AEF)$, we have $ABCDF \in \mathcal{I}(ABCDEF)$, so (∗∗) $ABCD \in \mathcal{I}(ABCDE)$, so $BD \in \mathcal{I}(ABCDEG)$ and $ACEG \in \mathcal{F}(ABCDEG)$.

Suppose $ABCD \in \mathcal{F}(ABCDEG)$, then $EG \in \mathcal{I}(ABCDEG)$, so $E \in \mathcal{I}(ABCDE)$, contradicting (∗∗).

□

5.2.3 Filters Generated by Preferential Relations

When we discuss $\prec$ on U, and $<_X$, $<$, $<'_X$, $<'$ for subsets of U, we implicitly mean the filters, ideals, etc. generated by μ on subsets of U, as discussed in Fact 5.2.10 (page 88), or a relation $\prec$ on U, as defined in Definition 5.2.4 (page 89).

We give some examples.

Example 5.2.1

(1) Let $\prec$ not be transitive.

Let $z \prec y \prec x$, then $\{x\} <_{\{x,y\}} \{y\}$, $\{y\} <_{\{y,z\}} \{z\}$, but $\{x\} \not<_{\{x,z\}} \{z\}$, as $\mu(\{x,z\}) = \{x,z\}$. However, $\{x\} <_{\{x,y,z\}} \{z\}$.

(2) Let $\prec$ be transitive.

In Case (1), add $z \prec x$, then $\{x\} \prec_{\{x,z\}} \{z\}$.

(3) Let $\prec$ again be transitive.

Consider $A = \{a\}$, $C = \{c\}$, $B = \{b_i : i < \omega\} \cup \{b\}$, and $b_i \prec a$, $c \prec b$, $b_{i+1} \prec b_i$. $\prec$ is transitive. Then $\mu(A \cup B) = \{b\}$, $\mu(B \cup C) = \{c\}$, $\mu(A \cup C) = \{a,c\}$. So $A <_{A \cup B} B$, $B <_{B \cup C} C$, but not $A <_{A \cup C} C$. However, $\mu(A \cup B \cup C) = C$, so $A <_{A \cup B \cup C} C$.

(4) Let $\prec$ again be transitive.

Let $X = \{x\} \cup \{x_i : i < \omega\}$, $Y = \{y\}$, $Z = \{z\} \cup \{z_i\}$, and $y \prec x$, $z_0 \prec y$, $x_0 \prec z$, so $z_0 \prec x$ by transitivity (and $x_j \prec x_i$ for $j > i$ etc.).

We have $\mu(X \cup Y) = \{y\}$, $\mu(Y \cup Z) = \{z\}$, $\mu(X \cup Z) = \emptyset$, so $X <_{X \cup Y} Y$, $Y <_{Y \cup Z} Z$, but $X \not<_{X \cup Z} Z$ and $X \not<_{X \cup Y \cup Z} Z$.

(5) Again, $\prec$ is transitive, in addition, $\prec$ is smooth.

Consider $X := \{x_2, x_3, x_4\}$, $Y := \{x_1, x_2, y\}$, $x_4 \prec x_2$, $y \prec x_3$, $y \prec x_1$ (the transitivity condition is empty).

Then $\mu(X) = \{x_3, x_4\}$, $\{x_3\} \in \mathcal{M}(X)$, $\{x_2\} \in \mathcal{I}(X)$, $\{x_2\} <'_X \{x_3\}$.

$\mu(Y) = \{x_2, y\}$, $\{x_2\} \in \mathcal{M}(Y)$, $\{x_1\} \in \mathcal{I}(Y)$, $\{x_1\} <'_Y \{x_2\}$.

Let $x_1, x_3 \in Z$, is $\{x_1\} <'_Z \{x_3\}$?

If $y \in Z$, $\{x_1\}, \{x_3\} \in \mathcal{I}(Z)$.

If $y \notin Z$, $\{x_1\}, \{x_3\} \in \mathcal{M}(Z)$.

So, in both cases, $\{x_1\} \not<'_Z \{x_3\}$, as they have the same size.

Example 5.2.2

$<$ is neither upward nor downward absolute. Intuitively, in a bigger set, formerly big sets might become small, conversely, in a smaller set, formerly small sets might become big.

Let $A, B \subseteq X \subseteq Y$. Then

(1) $A <_X B$ does not imply $A <_Y B$

(2) $A <_Y B$ does not imply $A <_X B$

(1): Let $Y := \{a, b, c\}$, $X := \{a, b\}$, $c \prec b \prec a$. Then $\{a\} <_X \{b\}$, but both $\{a\}, \{b\} \in \mathcal{I}(Y)$.

(2): Let $Y := \{a, b, c\}$, $X := \{a, c\}$, $c \prec b \prec a$, but NOT $c \prec a$. Then $\{a\} <_Y \{c\}$, but both $\{a\}, \{c\} \in \mathcal{M}(X)$.

For homogenousness, we chose violation in a comparatively smaller subset. As said above, this corresponds to the non-monotonicity idea, and, intuitively, going from a big set to a very small set, more things can happen. The smaller a subset, the less likely homogenousness is expected.

We then have a construction similar to defeasible inheritance as metatheory, so overall a coherent approach on object and meta level.

5.3 Application to Argumentation

5.3.1 Introduction

Abstract Description Argumentation is about putting certain objects together. The interested reader might compare this to the constructions in Section 3.5 (page 60) and in Chapter 3 (page 45).

There are three things to consider:

(1) the objects themselves, and their inner structure (if they have any) - this inner structure may be revealed successively, or be immediately present,

(2) rules about how to put them together,

(3) avoid certain results (contradictions) in the resulting pattern.

To help intuition, we picture as result of an argumentation, an inheritance network the agents can agree on.

This network may consist of strict and defeasible rules only, with no elements or sets it is applied to. Think of the argumentation going on when writing a book about medical diagnosis. This will not be about particular cases, but about strict and default rules. "Sympton x is usually a sign of illness y, but there are the following exceptions: ..." In addition, the network might contain cycles. There is nothing wrong with cycles. Mathematics is full of cycles, equivalences and their proofs. But consider also the following: We work in the set of adult land mammals. "Most elefants weigh more than 1 ton." "Most elements (i.e. adult land mammals) which weigh more than 1 ton are elefants." There is nothing in principle wrong with this either - except, in reality, we forgot perhaps about hippopotamus etc.

Arguments need not be contradictions to what exists already. They can be confirmations, elaborations, etc. For instance, we might have the default rule that birds fly, and clarify that penguins don't fly. This is not a contradiction, but an elaboration.

The structure of the objects Facts are either so simple that a dispute seems unreasonable. Or, they are a combination of basic facts and (default) rules, like, what I see through my microscope is really there, and not an artifact of some speck of dust on the lenses. For simplicity, facts will be basic, undisputable facts.

Expert opinion may be considered a default rule, where details stay unexplained, perhaps even unexplainable by the expert himself.

Rules (classical or defaults) have three aspects:

(1) the rule itself,

(2) the application of the rule,

(3) the result of the application of the rule.

Classical rules cannot be contested. We can contest their application, i.e. one of their prerequisites, or their result, and, consequently, their application. We can confirm their result by different means, likewise, their application.

Default rules are much more complicated, but not fundamentally different. Again, we can attack their application, by showing that one of the prerequisites does not hold, or, that we are in an (known) exceptional case. We can attack the conclusion, and, consequently, the rule, or its application. In particular, we may attack the conclusion, without attacking the application or the rule itself, by arguing that we are in a surprising exceptional case - and perhaps try to find a new set of exceptions. We can attack the default rule itself, as in the case of "normally, tigers are vegans". We can confirm a rule by confirming its conclusion, or adding a new rule, which gives the same result. We can elaborate a default rule, by adding an exception set, stating that all exceptions are known, and give the list of exceptions, etc. We can stop homogenousness (downward inheritance) e.g. for Quakers which are Republicans, we stop inheriting pacifism (or its opposite). This is not a contradiction to the default itself, but to the downward inheritance of the default (or to homogenousness) by meta-default, to be precise.

Obviously, the more we add (possible) properties to the objects (here default rules), the more we can attack, elaborate, confirm.

In the following section, we describe our general picture:

(1) there is an arbiter which checks for consistency, and directs the discussion,

(2) how to handle classical arguments and resulting contradictions,

(3) how to handle default arguments.

5.3.2   The Classical Part

We suppose there is an arbiter, whose role is to check consistency, and to authorise participants to speak.

If the arbiter detects an inconsistency, then he points out the "culprits", i.e. minimal inconsistent sets. As he detects inconsistencies immediately, the last argument will be in all those sets. The last argument need not be the problem, it might be one of the earlier arguments.

He asks all participants if they wish to retract one of the arguments involved in at least one minimal inconsistent set. (They have to agree unanimously on such retraction.) If there is no minimal inconsistent set left, the argumentation proceeds with the "cleaned" set of arguments, as if the inconsistency did not arise. Of course, arguments which were based on some of the retracted arguments are now left "hanging in the air", and may be open to new attacks.

If not, i.e. at least one minimal inconsistent set is left, the participants can defend (and attack) the arguments involved in those sets. The arbiter will chose the argument to be attacked/defended. See Example 5.3.1 (page 99) below. Suppose α is the argument chosen, then a defense will try to prove or argue for α, an attack will try to prove or argue for $\Diamond\neg\alpha$, i.e. it is possible or consistent that $\neg\alpha$. In particular, an attacker might try to prove $\bot$, or some other unlikely consequence of α (and some incontested β's), and he need not begin with some $\alpha \to \gamma$, it might be a more roundabout attack.

If at least one minimally inconsistent set is left with all elements defended, then there is a deadlock, and the arbiter declares failure.

Consider

Example 5.3.1

We argue semantically. Let $A := \{x, a\}$, $B := \{x, b\}$, $C := \{x, c\}$, $Y := \{a, b, c\}$. Let Y be the last set added. For A, B, C, the situation is symmetrical. Let $Z \neq Z'$ be A, B, or C, then $Y \cap Z \neq \emptyset$, but $Y \cap Z \cap Z' = \emptyset$, $Z \cap Z' = \{x\}$, etc. Moreover, $A \cap B \subseteq C$, etc. Thus, A and B together are an argument for C, etc., so they argue for each other, and there is no natural way to chose any of A, B, C to be attacked. Thus, it is at the discretion of the parties involved (or the arbiter) to chose the aim of any attack - apart from Y, which is not supported by any of A, B, C. Still, Y might in the end be the strongest argument.

We may add D, E, with $D := \{x, d\}$, $Y := \{a, b, c, d\}$ etc., the example may be extended to arbitrarily many sets.

At any moment, any argument can be attacked, not only if an inconsistency arises. We may continue an argumentation, even if not all minimally inconsistent subsets are treated as yet, but the arbiter has to keep track of them, and of the use of their elements. They and their consequences may still be questioned.

5.3.3 Defaults

The classical part of defaults We see defaults primarily not as rules, but as relatively complicated classical constructions, which we may see as objects for the moment. The default character is in applying those objects, not in the objects themselves.

We follow here the theory described in Chapter 11 of [GS16].

In our view, a (semantical) default $(X : Y)$ says:

(1) "most" elements of X are in Y,

(2) there may be exception sets X_1, X_2, etc. of X, where the elements are "mostly" not in Y (but $X_1 \cup X_2 \cup \ldots$ has to be a "small" subset of X),

(3) in addition, there may be a "very small" subset $X' \subseteq X$, which contains "surprise elements" (i.e. not previously known exceptions), which are not in Y,

(4) in addition, we may require that subsets of X "normally" behave in a homogenous way.

The notions of "most", "small" etc. are left open, a numerical interpretation suffices for the intuition. These notions are discussed in depth e.g. in [GS08f] and [GS10].

Introducing a default has to result in a (classically) consistent theory. E.g., it must not be the case that $\forall x \in X. x \notin Y$, this contradicts the first requirement about defaults (and any reasonable interpretation of "most").

The default part of defaults   This leads to a hierarchy as defined in Section 11.4.1 of [GS16]. We use the hierarchy to define the *use* of the defaults.

To use the standard example with birds, penguins, fly, we proceed as follows. Suppose we introduce a bird x into the discussion. We try to put x as low as possible in the hierarchy, i.e. into the set of birds, but not into any known exception set, and much less into any "surprise" set. Only (classical) inconsistency, as checked by the arbiter, may force us to climb higher. Thus, unless there is a contradiction, we let x fly.

Attacks against defaults and their conclusions   Classical rules are supposed to be always true. Thus, classical rules themselves cannot be attacked, and an attack against a classical conclusion has to be an attack against one of its prerequisites.

Attacks against defaults can be attacks against

(1) the rule itself,

(2) one of the prerequisites,

(3) membership in or not in one of the exception sets,

(4) membership in or not in the surprise set,

(5) perhaps even the notions of size involved,

(6) etc.

Each component of a default rule may be attacked.

5.3.4   Comments

We assume that there is no fundamental difference between facts and conclusions: Usually, we were told facts, remember facts, have read facts, observed facts (perhaps with the help of a telescope etc.). These things can go wrong. Situations where things are obvious, and no error seems humanly possible, will not be contradicted.

Auxiliary elements We now introduce some auxiliary elements which may help in the argumentation.

(1) "I agree."

This makes an error in this aspect less likely, as both parties agree - but still possible!

(2) "I confirm."

I am very certain about this aspect.

(3) Expert knowledge:

Expert knowledge and its conclusions act as "black box defaults", which the expert himself may be unable to analyse. Other experts (in the same field) will share the conclusion. (This is simplified, of course.)

(One way to contest an expert's conclusion is to point out that he neglected an aspect of the situation, which is outside his expertise. His "language of reasoning" is too poor for the situation.)

(4) The arbiter may ask questions.

Examples of attacks

(1) Defaults:

Normally, there is a bus line number 1 running every 10 minutes between 10 and 11 in the morning.

Attack: No, the conclusion is wrong.

Question: Why?

Elaboration:

(1.1) No, the default is wrong (e.g.: it is line number 2 running every 10 minutes).

(1.2) Yes, but this is not homogenous, i.e. does not break down to subsets, and we know more. (For instance, we know that today is Tuesday or Wednesday, and it runs that often only Monday, Thursday, Friday, Saturday, Sunday - but we do not know this, only that is does not apply to all days of the week.)

(1.3) Yes, but today is an exception, and we know this. (e.g., we know that today is Tuesday, and we know that Tuesday is an exception.) (In addition, there might be exceptional Tuesdays, Christmas market day, etc. ...)

(1.4) Yes, but I do not know why this is an exception. (This is a surprise case, I know about different days, but today should not be an exception, still I was just informed that it does not hold today.) We do not attack the default, nor the applicability - but agree that it fails here.

(2) Classical conclusions:

From A and B, C follows classically.

Attack: C does not hold.

Question: Why?

Elaboration:

(2.1) A does not hold or B does not hold, but I do not know which.

(2.2) A does not hold.

(2.3) B does not hold.

(2.4) A does not hold, and B does not hold.

(3) Fact: A holds.

Attacks: No, A does not hold.

Question: Why?

Elaboration:

(3.1) You remember incorrectly.

(3.2) You did not observe well.

(3.3) Your observation tools do not work.

(3.4) You were told something wrong.

(3.5) etc.

(4) Expert knowledge, expert concludes that A.

Attack: A does not hold.

Question: Why?

Elaboration:

The situation involves aspects where you are not an expert. It is beyond your language. (Of course, the expert can ask for elaboration ....)

We did not treat here:

(1) when it is necessary to remember not only the result of an argument, but also the way it was reached, (compare this to Chapter 2 (page 7), where we remembered whether we approximated from above or from below.

(2) the usual "dirty tricks" of political argumentation like:

- changing focus,

- attack unimportant details, etc.

5.3.5   Various other Applications

(1) Inheritance:

We refer the reader to Section 5.8 of [Sch18b], which contains a detailed discussion, and just add some remarks.

Apart from any formal reasons, there might be philosophical or even pragmatic arguments to choese one way or the other for

  (1.1) reference classes: a user might think boolean combinations of reference classes natural, or surprising, this may influence our decision,

  (1.2) we may consider, beyond specificity, length of path, in particular if we have proof that the arrows in the path describe size relations (see Fact 5.2.9 (page 88)),

  (1.3) the decision for extensions or direct scepticism might depend on whether the problem reflects a lack of information, or rather too much contradictory information.

(2) We developed similar ideas in Chapter 11 of [GS16], and refer the reader there.

(3) Analogical reasoning: See Chapter 4 (page 69).

Chapter 6

A Reliability Theory of Truth

6.1   Introduction: Motivation, Example and Basic Idea

6.1.1   Motivation

Our motivation is not to detect inconsistencies in present theories of truth, and how to remedy them, but to separate truth from falsity in a flood of information.

The problem is acerbated by a strategy to destroy truth as an important criterion in political and other discussions. Jonathan Rauch's important book [Rau21], discusses these efforts in detail, culminating perhaps in Steve Bannon's "…. flood the zone with shit." (Of course, other countries' behaviour is not better, see China, Russia, etc.)

Similar problems appear in the myths surrounding the Covid pandemic, where rumours without the slightest factual foundation abound.

Thus, we think, it is very important to have a theory that tries to help distinguish facts from myths (or worse), leading perhaps even to algorithms which help to sort today's flood of (dis-)information.

6.1.2   Example and Basic Idea

We continue with a simple example.

Example 6.1.1

Suppose we want to know the temperature in a room. We have four thermometers, and no other way to know the temperature.

T_1 one says 20 C, T_2 says 19 C, T_3 21 C, and T_4 says 30 C. Thus, T_4 reports an exceptional value, and we doubt its reliability.

How do we model this? A simple idea is as follows: Each T_i is given a reliability $\rho(T_i)$ between 0 (totally unreliable) and 1 (totally reliable). At the beginning, each $\rho(T_i)$ is a neutral value, say 0.5. We now calculate the mean value, $90/4 = 22.5$. As we start with equal reliability, each T_i is given the same weight 0.5. We see now that T_4 is exceptional, and adjust reliabilities, e.g. $\rho(T_4) = 1/3$, and $\rho(T_i) = 2/3$ for the other i. If, in the next moment, all T_i give again the same data, we will adjust the mean value, by counting the values for T_1 to T_3 twice, the value for T_4 once, and divide by 7, resulting in $(120 + 30)/7 = 21.43$. Etc.

This is our basic idea. It seems a reasonable way to treat contradictory numerical information, and some variant is probably used in many "real life" situations where we need some information, cannot trust absolutely any single source, but need the information, e.g. to act.

We do not doubt that there is some "real" temperature of the room, but this is irrelevant, as we cannot know it. We have to do with what we know, but are aware that additional information might lead us to revise our estimate.

There are a number of ways to elaborate, modify, and apply to different situations.

(1) First, we work here with numerical values, both the data and reliabilities are real numbers, so is the mean value. We want to do more. We want to work with totally ordered sets instead of reals, at least for the data, then with partial orders complete under sup and inf, perhaps complement, and, finally, with arbitrary partial orders. So, we have to try to adapt our data and operations in some way or the other to the limited possibilities of the structure at hand. We do not claim that our suggestions are the only or best ways to proceed, the "right" way may also depend on the situation. We note here which operations on the reals we use, and refer the reader to Chapter 2 (page 7) for adaptations to weaker structures.

(2) Then, even the numerical case need not have a unique solution. For instance, when calculating the mean value, we might give less (or more!) weight to exceptional values, without considering reliabilities. This could be done, e.g., by calculating first the usual mean, and then "pull" the exceptional values closer to the mean, and calculate the mean value again. In above example, once we calculated the mean value, 22.5, we note the exceptional difference between 22.5 and 30, modify 30 to 27.5, and calculate the mean of $\{19, 20, 21, 27.5\}$, etc.

We will not follow all such possibilies, but use abstract functions, here "mean", in Example 6.1.1 (page 105) we have $mean(\{19, 20, 21, 30\}) = 22.5$. We will, however, indicate the translation of all variants discussed in detail to less rich domains.

(3) The communication channels may have a reliability, too. So the message which arrives has a combined reliability of the agent's reliability and the messages reliability. How do we calculate this combination, and conversely, when adjusting the overall reliability, how do we adjust the individual ones of agent and channel?

(4) In our example, we have one agent which listens and calculates, the other agents measure and send values to the "central" agent. Moreover, the sending is synchronised. We might also have situations where the measuring and calculating agents are the same, and the messages are broadcast.

(5) The measuring agents might have an estimate about their own reliability, and communicate this with their data. E.g., in above Example, a thermometer may have different precisions for different temperature ranges, e.g., very good from 10 to 30 degrees C, from 0 to 10, and 30 to 40 not so good, etc.

(6) The agents may have opinions about the reliabilities of other agents, think of politicians who consider each others crooks, so the data may not only be "facts", but also reliabilities of other agents.

(7) The history should perhaps be preserved beyond the individual reliabilities. Suppose we measured above temperature repeatedly (and the temperature is supposed to be constant), then, in order to calculate the mean over time, we need to memorize past results or mean values in some way.

Agents may be people, devices like thermometers, theories, etc. Sometimes, it is more adequate to see reliability as degree of competence, for instance for moral questions. Messages may be numbers, but also statements, like the earth is flat. The formal treatment of such cases is discussed in Chapter 2 (page 7).

A human agent may be a good chemist, but a poor mathematician, so his reliability varies with the subject. For simplicity, we treat this agent as two diffent agents, A-Chemist, A-Mathematician, etc.

Philosophical theories of truth are often mainly about contradictions, in the tradition of the liar paradox (see Chapter 7 (page 117)). Our approach is very different. We create on the fly new truth values, they do not stand for "true" and "false", but for more or less reliable, and whenever we need a new value of reliability, we create it.

We will say more about the comparison of out idea to other theories of truth below.

Note that a theory and corresponding algorithms to help decide between reliable und unreliable information are particularly important in the present flood of misinformation. Similarly, stock markets need good algorithms to differentiate between changes based on underlying facts and mere contagion of behaviour between agents.

6.2   In more Detail

We now address above points.

6.2.1   The Basic Scenario With some Features Added

6.2.1.1   The Basic Scenario with History Added

Agents A_i send numerical values r_i to a central evaluation agent E. These are the only messages sent. Each agent A_i has a real value reliability $\rho_i \in [0,1]$ which is determined by E. 0 stands for total unreliability, 1 for total reliability. At the outset, each ρ_i will have the neutral value 0.5. (The agents will not know their reliabilities, nor those of other agents.)

We will indicate the (additional for Variant 2 upward and the following sections) operations needed, and which will have to be adapted in non-numerical settings.

At a given time, agents A_i send their r_i to E. This is done synchonously. Once all r_i are received by E :

(1) Variant 1:

   E calculates the mean (average) m of all r_i. The closer the individual r_i is to m, the more reliable r_i is considered, the better ρ_i. More precisely: Let δ_i be the distance from r_i to m, and let δ be the mean of all δ_i. The better δ_i is in comparison to δ, the more reliable A_i seems to be.

   We calculate the new ρ_i by a suitable function: $\rho_i' := f_\rho(\rho_i, \delta, \delta_i)$. If δ_i is better than δ, we increase ρ_i, if not, we decrease ρ_i. The precise details will not matter, and depend also on the context. (We might, e.g., have a minimal threshold of discrepancy, below which we do nothing.)

   Operations:

   (1.1) m (mean value) of the r_i

   (1.2) δ_i = distance between m and r_i

   (1.3) δ = mean value of all δ_i

   (1.4) adjusting ρ_i using δ, δ_i, and old ρ_i

(2) Variant 2:

   E uses the old ρ_i already to give different weight to the r_i. E.g., if ρ_i is twice as good as ρ_j, we may count r_i twice (and r_j once), to give it more weight, as A_j has a "bad reputation", and A_i a good one. The rest is the same as in Variant 1.

   Operations:

   adjust r_i using old ρ_i

(3) Variant 3:

   As in Variant 1 or 2, but we assume we have already earlier measurements of the same entity (assumed constant), so we have already an "old" m, which summarizes the old data, the history. The "inertia" t of the old m should express the number of r_i which went into the calculation of the old m.

Thus, e.g., we enter the old m t times, just as we would enter t new r_i. We may modify, e.g. give the old m more or less weight, etc.

In the same way, we may give the old $\rho_i's$ more or less weight.

This way, we may also treat asynchronous arrival of messages from different agents. Some precaution against receiving repeated messages from the same agent might be necessary.

Operations:

(3.1) purely administrative: count numer of r_i

(3.2) multiply old m by t

(4) Variant 4:

For this variant, we need to put the ρ_i in relation to the values r_i (and m). Suppose, e.g., that an agent A_i's r_i should be within 10% of m if ρ_i is 0.9. The better ρ_i, the more r_i should be close to m. We might then decide to decrease ρ_i, if δ_i is too big for ρ_i, etc. Details, again, are not important, the relation of ρ_i to δ_i, and thus between reliability and data, is important.

Operations:

put ρ_i in relation to a difference between m and r_i

6.2.1.2   Hypotheses About own Reliability and Reliability of Communication Channels

An agent may have a hypothesis about the reliability of his own message. E.g., a human being might caution that his expertise is not very good in a certain field, or that he feels very confident. A thermometer may have a temperature range where it is very precise, and outside this range it may be less so - and it may "know" about it. Thus, the message has two parts, data, and presumed reliability, say $\in_i$. Consider above Variant 4. If δ_i is not too big in relation to $\in_i$, E may renounce on decreasing ρ_i, as the agent was aware of the problem. In addition, E may give less weight to r_i, see Variant 2.

Communication channels may have a reliability, too, say ρc_i. The value r_i has now combined reliability of ρ_i and ρc_i. The simplest way to combine them might be multiplication, and it should probably not be bigger than $min\{\rho_i, \rho c_i\}$. Conversely, if we want to modify the combined reliability, we have to decide how to modify both parts. Multiplication by a common factor seems a simple way to proceed.

Operations:

(1) (serial) combination of two reliabilities, here ρ_i and ρc_i

(2) conversely, break down a modification of a combination of two reliabilities to a modification of the individual reliabilities (this should be an inverse operation to the first operation here)

6.2.2   Broadcasting and Messages About Reliability of other Agents

6.2.2.1   Broadcasting

One problem with broadcasting (anyone may send messages to anyone) is that it may lead to contradictory or self-supporting cycles. E.g. agent A sends a message to B whereupon B sends a message to A amplifying A's message and so forth. A need not see that it is just A's own message coming back stronger. Adding history to the messages solves the problem. Suppose A sends the message $\langle A, r \rangle$, expressing that it is a message from A, and B sends the message expressing that it is a reply $\langle A, r, B, r' \rangle$, then A sees that the message originated from A, and will not send it again. Likewise, B might not send it to A, as A "saw" it already.

If E listens in to all messages, E can detect such cycles, and react accordingly, e.g. contract the whole group to a single agent, or neglecting the whole group, if it is infighting.

The problem of oscillations is a common one, and it might be interesting to see how the brain avoids them.

Operations:

Note that we do not need any new operations as the new elements are about control.

6.2.2.2   Messages about Reliabilities of other Agents

Messages about reliabilities of other agents may easily be destructive (as can be seen in politics!). Let $\in_{i,j}$ be a message from agent i about the reliability of agent j.

Suppose A_i sends $\in_{i,j}$ to E.

First, how much weight should E give to $\in_{i,j}$? It should probably not be totally neglected, but history (which E should store) should matter. E.g., if A_i and A_j support each other positively, this might (but need not) be cronyism, if they do so negatively (e.g. A_i and A_j say that the other is unreliable), it might be a case of infighting.

The problem is easy to see, but there is probably no general solution, only answers to particular cases.

If we allow such messages to be broadcast, they might end in positive or negative cycles, which an "umpire" should detect and prevent.

Operations:

Again, we do not need any new operations as the new elements are about control.

6.2.2.3   Operations On the Contents of the Messages

We do not discuss operations on the contents of the messages beyond this short remark. E.g., the reliability of a value in the interval $[10, 20]$ should be at least as good as individual reliabilities for $x \in [10, 20]$, the same applies for the reliability of $\phi \vee \psi$ in relation to the reliabilities for ϕ and ψ.

6.3   Discussion

Remark 6.3.1

The following extensions seem possible:

- Actions and animals:

  We can apply similar reasoning to actions. The action of a monkey (the agent) which sees a lion and climbs a tree to safety is "true", or, better, adequate.

- Values:

  Values, obligations, "natural laws" (in the sense of philosophy of law) are subjective. Still, some influences are known, and we can try to peel them off. Religion, politics, personal history, influence our ideas about values. One can try to find the "common" and "reasonable" core of them. For instance, religious extremism tends to produce ruthless value systems, so we might consider religious extremists as less reliable about values.

Our approach is very pragmatic, a method, and takes its intuition from e.g. physics, where a theory is considered true - but revisably so! - when there is "sufficient" confirmation, by experiments, support from other theories, etc.

Automatic trading in financial markets has to consider some aspects of our ideas: one should caution against excessive feedback, as it might generate unfounded fluctuations.

Many human efforts are about establishing reliability of humans or devices. An egineer or physician has to undergo exams to assure that he is competent, a bridge has to meet construction standards, etc. All this is not infallible, experts make mistakes, new, unknown possibilities of failure may appear - we just try to do our best.

Our ideas are examples how it can be done, but no definite solutions. The exact choice is perhaps not so important, as long as there is a process of permanent adjustment. This process has proven extremely fruitful in science, and deserves to be seen as a powerful method, if not to find truth, at least to find "sufficient" information.

From an epistemological point of view, our position is that of "naturalistic episte-mology", and we need not decide between "foundationalism" and "coherentism", the interval $[0, 1]$ has enough space to maneuvre between more and less foundational information. See e.g. [Sta17c].

Our approach has some similarities with the utility approach, see the discussion in [BB11], the chapter on utility. An assumption, though false, can be useful: if you think a lion is outside, and keep the door closed, this is useful, even if, in fact, it is a tiger which is outside. "A lion is outside" is false, but sufficiently true. We think that this shows again that truth should not be seen as something absolute, but as something we can at best approximate; and, conversely, that it is not necessary to know "absolute truth". We go beyond utility, as improvement is implicit in our approach. Of course, approximation may only be an illusion generated by the fact that we develop theories which seem to fit better and better, but whether we approach reality and truth, or, on the contrary, move away from reality and truth, we cannot know.

There are many things we did not consider, e.g. if more complicated, strongly connected, structures have stronger inertia against adjustment.

We use meta-information (reliability - which, importantly, is not binary, not just true/false) to avoid mistakes, in our example of measurements. We further use control information to detect cycles and group behaviour, to avoid further mistakes. Of course, this is all very primitive, and further elaborations are possible and necessary, sometimes depending on the type and environment of the data. The question is whether the method is adequate, there are no completeness and correctness properties to be discovered - it is about methods, not logics, there are no axioms etc.

We have an example structure which handles these problems very well: our brain. Attacks, negative values of reliability, correspond to inhibitory synapses, positive values, support, to excitatory synapses. Complex, connected structures with loops are created all the time without uncontrolled feedback. It is perhaps not sufficiently clear how this works, but it must work! (The "matching inhibition" mechanism seems to be a candidate. See also [OL09] for a discussion of the "cooperation" of excitatory and inhibitory inputs of a neuron. E.g., excitation may be followed closely by inhibition, thus explaining the suppression of such feedback. The author is indebted to Ch. von der Malsburg, FIAS, for these hints.) Our theories about the world survive some attack (inertia), until "enough is enough", and we switch emphasis. The brain's mechanisms for attention can handle this.

6.4   Philosophical Background

6.4.1   The Coherence and Correspondence Theories of Truth

See [Sta17a] for an overview for the coherence theory, and [Sta17b] for an overview for the correspondence theory. The latter contains an extensive bibliography, and we refer the reader there for more details on the correspondence theory.

We think that the criticisms of the coherence theory of truth are peripheral, but the criticism of the correspondence theory of truth is fundamental.

The criticism of the correspondence theory, that we have no direct access to reality, and have to do with our limitations in observing and thinking, seems fundamental to the author. The discussion whether there are some "correct" theories our brains are unable to formulate, is taken seriously by physicists, likewise the discussion, whether e.g. Quarks are real, or only helpful "images" to understand reality, was taken very seriously. E.g. Gell Mann was longtime undecided about it, and people perhaps just got used to them. We don't know what reality is, and it seems we will never know. See also discussions in neurophilosophy, [Sta17d] for a general introduction.

On the other side, two main criticisms of the coherence theory can be easily countered, in our opinion. See e.g. [Rus07] and [Tha07] for objections to coherence theory. Russell's objection, that ϕ and $\neg\phi$ may both be consistent with a given theory, shows just that "consistency" is the wrong interpretation of "coherence", and it also leaves open the question which logic we work in. The objection that the background theory against which we check coherence is undefined, can be countered with a simple argument: Everything. In "reality", of course, this is not the case. If we have a difficult physical problem, we will not ask our baker, and even if he has an opinion, we will not give it much consideration. Sources of information are assessed, and only "good" sources (for the problem at hand!) will be considered. (Thus, we also avoid the postmodernist trap: there are standards of "normal reasoning" whose values have been shown in unbiased everyday life, and against which standards of every society have to be compared. No hope for the political crackpots here!)

Our approach will be a variant of the coherence theory, related ideas were also expressed by [Hem35] and [Neu83].

We can see our approach in the tradition of relinquishing absoluteness:

- The introduction of axiom systems made truth relative to axioms.

- Nonmonotonic reasoning allowed for exceptions.

- Our approach treats uncertainty of information, and our potential inability to know reality.

6.4.2 A Short Comparison of Our Approach to Other Theories

(1) Our approach is not about discovery, only about evaluating information.

(2) In contrast to many philosophical theories of truth, we do not treat paradoxa, as done e.g. in [Kri75] or [BS17], we assume statements to be "naive" and free from semantic problems.

We do treat cycles too, but they are simpler, and we take care not to go through them repeatedly. In addition, our structures are assumed to be finite.

(3) On the philosophical side, we are probably closest to the discourse theory of the Frankfurt School, in particular to the work by J. Habermas and K. O. Apel (as we discovered by chance!), see e.g. [Wik18b], [Sta18b], [Hab73], [Hab90], [Hab96], [Hab01], [Hab03].

Importantly, they treat with the basically same methods problems of truth and ethics, see our Remark 6.3.1 (page 111) below.

We see three differences with their approach.

(3.1) A minor difference: We also consider objects like thermometers as agents, not only human beings, thus eliminating some of the subjectivity.

(3.2) A major difference: We use feedback to modify reliability of agents and messages. Thus, the $\forall$-quantifier over participating agents in the Frankfurt School is attenuated to those considered reliable.

(3.3) Conversely, their discourse theory is, of course, much more developed than our approach.

Thus, an integration of both approaches seems promising.

(4) Articles on trust, like [BBHLL10] or [BP12], treat different, more subtle, and perhaps less fundamental, problems. A detailed overview over trust systems is given in [SS05].

We concentrate on logics, cycles, and composition of values by concatenation. Still, our approach is in methods, but not in motivation, perhaps closer to the basic ideas of trust systems, than to those of theories of truth, which often concentrate on paradoxa.

Articles on trust will often describe interesting ideas about details of coding, e.g. [BP12] describes how to code a set of numerical values by an interval (or, equivalently, two values).

(5) Basic argumentation systems, see e.g. [Dun95], will not distinguish between arguments of different quality. Argumentation systems with preferences, see e.g. [MP13], may do so, but they do not seem to propagate conflict and confirmation backwards to the source of arguments, which is an essential part of our approach. This backward propagation also seems a core part of any truth theory in our spirit. Such theories have to be able to learn from past errors and successes.

(6) Is this a Theory of Truth?

The author thinks that, yes, though we hardly mentioned truth in the text.

Modern physics are perhaps the best attempt to find out what "reality" is, what "truly holds". We had the development of physics in mind, reliability of experiments, measurements, coherence of theories (forward and backward influence of reliabilities), reputation of certain physicists, predictions, etc. Of course, the present text is only a very rough sketch, we see it as a first attempt, providing some highly flexible ingredients for a more complete theory in this spirit.

Chapter 7

Remarks on Yablo's Paradox

7.1   Introduction

Unless stated otherwise, we work in propositional logic, with disjunctive normal forms, i.e. formulas of the type $\bigvee \bigwedge$. Formulas may, however, be infinite.

7.1.1   Overview

After some definitions, we show in Section 7.2 (page 122) that a conjecture in [RRM13] is wrong.

In Section 7.3 (page 131), we discuss basic contradictions, cells, and give an example of basic reasoning about contradictory sequences, see Section 7.4.2 (page 155).

Section 7.3.2.3 (page 142) sees are detailed analysis of Yablo's construction, some aspects of his construction are hidden behind its elegance. This leads to the concepts of "head", "knee", and "foot", and then to "saw blades" in Section 7.6 (page 168). Section 7.5 (page 161) generalizes Yablo's construction to arbitrary formulas of the type $\bigvee \bigwedge$ (disjunctive normal forms), and offers a number of easy variations of such structures by modifying the order of the graph.

Section 7.6 (page 168) uses our idea of finer analysis of contradictory cells to build somewhat different structures - though the distinction is blurred by the necessarily recursive construction of contradictions.

We do not go in a straight line for the representation problem, but rather collect some ideas. We hope they are useful building blocks for a solution of the representation problem.

The author of the present text did not study the literature systematically. So, if some examples are already discussed elsewhere, the author would ask to be excused for not quoting previous work.

7.1.2   Basic Definitions and Results

We start with some notation and a trivial fact:

Definition 7.1.1

(1) $Inc(\phi)$ will stand for (classical propositional) inconsistency of ϕ

(2) $Cont(\phi)$ for "contradictory", i.e. $Inc(\phi) \wedge Inc(\neg\phi)$

    (by abuse of language, $\wedge$ etc. will be used in object and meta language).

We then have the trivial result

Fact 7.1.1

(1) $Cont(\phi)$ iff $Cont(\neg\phi)$

(2) $Cont(\phi \wedge \psi)$ iff $Inc(\phi \wedge \psi) \wedge Inc(\neg(\phi \wedge \psi))$.

$Inc(\neg(\phi \wedge \psi))$ iff $Inc(\neg\phi \vee \neg\psi)$ iff $Inc(\neg\phi) \wedge Inc(\neg\psi)$

$Inc(\phi) \Rightarrow Inc(\phi \wedge \psi)$, but (obviously) $Inc(\phi \wedge \psi) \wedge \neg Inc(\phi) \wedge \neg Inc(\psi)$ is possible.

Thus, $Cont(\phi) \wedge Inc(\neg\psi) \Rightarrow Cont(\phi \wedge \psi)$, but neither $Cont(\phi)$ nor $Cont(\psi)$ follow from $Cont(\phi \wedge \psi)$.

In particular, $Cont(\phi) \wedge$ TRUE $\Leftrightarrow Cont(\phi \wedge TRUE)$.

(3) $Cont(\phi \vee \psi)$ iff $Inc(\phi \vee \psi) \wedge Inc(\neg(\phi \vee \psi))$.

$Inc(\phi \vee \psi)$ iff $Inc(\phi) \wedge Inc(\psi)$,

$Inc(\neg(\phi \vee \psi))$ iff $Inc(\neg\phi \wedge \neg\psi)$, so $Inc(\neg\phi) \Rightarrow Inc(\neg(\phi \vee \psi))$.

Thus, $Cont(\phi) \wedge Inc(\psi) \Rightarrow Cont(\phi \vee \psi)$.

In particular, $Cont(\phi) \wedge$ FALSE $\Leftrightarrow Cont(\phi \vee FALSE)$

(4) Consequently, adding or eliminating a branch evaluating to TRUE with $\wedge$ will not change the "contradictory" status, neither will adding a branch evaluating to FALSE with $\vee$.

This is important for simplifications of a diagram.

Definition 7.1.2 (page 119), Definition 7.1.3 (page 120), and Definition 7.1.4 (page 121), are taken mostly from [RRM13].

Definition 7.1.2

(1) Given a (directed or not) graph G, $V(G)$ will denote its set of vertices, $E(G)$ its set of edges. In a directed graph, $xy \in E(G)$ will denote an arrow from x to y, which we also write $x \rightarrow y$, if G is not directed, just a line from x to y.

We often use x, y, or X, Y, etc. for vertices.

(2) A graph G is called transitive iff $xy, yz \in E(G)$ implies $xz \in E(G)$.

(3) Given two directed graphs G and H, a homomorphism from G to H is a function $f : V(G) \rightarrow V(H)$ such that, if $xy \in E(G)$, then $f(x)f(y) \in E(H)$.

(4) Given a directed graph G, the underlying undirected graph is defined as follows: $V(U(G)) := V(G)$, $xy \in E(U(G))$ iff $xy \in E(G)$ or $yx \in E(G)$), i.e., we forget the orientation of the edges. Conversely, G is called an orientation of $U(G)$.

(5) S, etc. will denote the set of propositional variables of some propositional language $\mathcal{L}$, S^+, etc. the set of its formulas. $\top$ and $\bot$ will be part of the formulas.

(6) Given $\mathcal{L}$, v will be a valuation, defined on S, and extended to S^+ as usual - the values will be $\{0, 1\}$, $\{\top, \bot\}$, or so. $[s]_v$, $[\alpha]_v$ will denote the valuation of $s \in S$, $\alpha \in S^+$, etc. When the context is clear, we might omit the index v.

(7) d etc. will be a denotation assignment, or simply denotation, a function from S to S^+.

Note that d need not have any logical meaning, it is an arbitrary function.

We sometimes abbreviate, e.g. $d(x) = y \wedge \neg z$ will be written $x = y \wedge \neg z$, etc.

The arrows in the graph $G_{S,d}$ below will point to the variables in $d(s)$, not to $d(s)$ or so.

(8) A valuation v is acceptable on S relative to d, iff for all $s \in S$ $[s]_v = [d(s)]_v$, i.e. iff $[s \leftrightarrow d(s)]_v = \top$. (When S and d are fixed, we just say that v is acceptable.)

(9) A system (S, d) is called paradoxical iff there is no v acceptable for S, d.

(10) Given S, d, we define $G_{S,d}$ as follows: $V(G_{S,d}) := S$, $ss' \in E(G_{S,d})$ iff $s' \in S$ occurs in $d(s)$.

If there are no arrows originating in s, then $d(s)$ is equivalent to $\bot$ or $\top$.

For clarity, one might write $d(s)$ next to s in $G_{S,d}$, but this would further complicate the graphs. But, of course, $d(s)$ is essential for the comprehension.

(11) A directed graph G is dangerous iff there is a paradoxical system (S, d), such that G is isomorphic to $G_{S,d}$.

Comment 7.1.1

Note that (9) and (11) give very different representation problems, (11) offers much more freedom, as justified by Fact 7.1.1 (page 118) (4).

In (11), we are given only the variables ocurring in $d(x)$, and may build up any formula with them. Fact 7.1.1 (page 118) (4) may thus offer ways to simplify a problem, e.g. by interpreting $x \to a$ as $a \vee \neg a$, $a \wedge \neg a$, etc., whenever it is possible to give classical truth values.

This suggests a strategy of pre-processing: if we want to examine whether a structure has a contradictory interpretation, chose suitable classical truth values whenever possible (e.g. if the structure below x is finite, has finite depth, is a tree, etc.) to simplify the problem. Of course, finding possible classical truth values is dual to finding contradictory truth value, $Cont(x)$ above.

Definition 7.1.3

Let G be a directed graph, $x, x' \in V(G)$.

(1) x' is a successor of x iff $xx' \in E(G)$.

$succ(x) := \{x' : x' \text{ is a successor of } x\}$,

(2) Call x' downward from x iff there is a path from x to x', i.e. x' is in the transitive closure of the succ operator.

(3) Let $[x \rightarrow]$ be the subgraph of G generated by $\{x\} \cup \{x' : x' \text{ is downward from } x\}$, i.e. $V([x \rightarrow]) := \{x\} \cup \{x' : x' \text{ is downward from } x\}$, and $x' \rightarrow x'' \in E([x \rightarrow])$ iff $x', x'' \in V([x \rightarrow])$, and $x'x'' \in E(G)$.

Definition 7.1.4

For easier reference, we define the Yablo structure, see e.g. [RRM13].

Let $V(G) := (Y_i : i < \omega)$, $E(G) := \{Y_i Y_j : i, j < \omega, i < j\}$, and $d(Y_i) := \bigwedge\{\neg Y_j : i < j\}$.

$(Y_i \in S$ for a suitable language.)

We see immediately the following simple, but very important fact:

Fact 7.1.2

(1) The logic as used in Yablo's construction is not compact.

(2) It is impossible to construct a Yablo-like structure with classical logic.

Proof

(1) Trivial.

(Take $\{\bigvee\{\phi_i : i \in \omega\}\} \cup \{\neg\phi_i : i \in \omega\}$. This is obviously inconsistent, but no finite subset is.).

(2) Take an acyclic graph, and interpret it as in Yablo's construction. Wlog., we may assume the graph is connected. Suppose it shows that x_0 cannot be given a truth value. Then the set of formulas showing this does not have a model, so it is inconsistent. If the formulas were classical, it would have a finite, inconsisten subset, Φ. Define the depth of a formula as the shortest path from x_0 to this formula. There is a (finite) n such that all formulas in Φ have depth $\leq n$. Give all formulas of depth n (arbitrary) truth values, and work upwards using truth functions. As the graph is acyclic, this is possible. Finally, x_0 has a truth value.

Thus, we need the infinite $\bigwedge / \bigvee$.

□

Remark 7.1.3

By the same argument as in the second half of (2) above, we see that we need infinite descending chains to obtain Yablo's Paradox.

Notation 7.1.1

(1) As shorthand, we will sometimes use:

   $x+$ will mean that x is true, likewise $x-$ that x is false, $x\pm$ that x is contradictory, i.e. it cannot have a truth value in the structure considered.

(2) $x \to y$ will mean that y occurs positively in $d(x)$, e.g. $d(x) = y \wedge \neg z$, in the same example we would write $x \not\to z$.

(3) $x \Rightarrow_\pm y$ (or $x \to_\pm y$) stands for $d(x) = y \wedge \neg y$.

We begin with some trivialities, just to remind the reader.

Remark 7.1.4

(1) When we construct a structure, we may have e.g. the choice of constructing two or three branches. If we construct an example (and not all cases), we can chose as we like - from the outside so to say. Once we did chose a structure, we are not free any more, we have to follow all branches in a $\forall$ situation - inside, we are not free any more.

(2) When we want to show that $\phi = \phi' \vee \phi''$ is contradictory, we have to show that both ϕ' and ϕ'' are contradictory.

   When we want to show that $\phi = \phi' \wedge \phi''$ is contradictory, it suffices to show that one of ϕ' or ϕ'' is contradictory (or both together).

(3) If the structure has no infinite descending chains, then there is a consistent valuation:

   We may give arbitrary values to the bottom elements, and calculate upwards, using the truth functions.

(4) We have to show that for some node x in the structure, assigning TRUE to x leads to a contradiction, and assigning FALSE to x will also lead to a contradiction.

(5) We do not need constants $\top, \bot$.

   Instead of assigning $\top$ to some node y, we may introduce a new node y' and define $y := y' \vee \neg y'$, similarly for $\bot$.

7.2   Comments on Rabern et al., [RRM13]

7.2.1   Introduction

This section is a footnote to [RRM13]. [RRM13] is perhaps best described as a graph theoretical analysis of Yablo's construction, see [Yab82]. We continue this work.

To make the present paper self-contained, we repeat the definitions of [RRM13]. To keep it short, we do not repeat ideas and motivations of [RRM13]. Thus, the reader should probably be familiar with or have a copy of [RRM13] ready.

All graphs etc. considered will be assumed to be cycle-free, unless said otherwise.

7.2.1.1 Overview

(1) Section 7.1.2 (page 118) contains most of the definitions we use, many are taken from [RRM13].

(2) In Section 7.2.2 (page 124), we show that conjecture 15 in [RRM13] is wrong. This conjecture says that a directed graph G is dangerous iff every homomorphic image of G is dangerous. (The definitions are given in Definition 7.1.2 (page 119), (3) and (11).)

  To show that the conjecture is wrong, we modify the Yablo construction, see Definition 7.1.4 (page 121), slightly in Example 7.2.1 (page 125), illustrated in Diagram 7.2.1 (page 127), show that it is still dangerous in Fact 7.2.3 (page 126), and collaps it to a homomorphic image in Example 7.2.2 (page 126). This homomorphic image is not dangerous, as shown in Fact 7.2.2 (page 124).

(3) In Section 7.2.3 (page 128), we discuss implications of Theorem 24 in [RRM13] - see the paragraph immediately after the proof of the theorem in [RRM13]. This theorem states that an undirected graph G has a dangerous orientation iff it contains a cycle. (See Definition 7.1.2 (page 119) (4) for orientation.)

  We show that for any simply connected directed graph G - i.e., in the underlying undirected graph $U(G)$, from any two vertices X, Y, there is at most one path from X to Y, see Definition 7.2.4 (page 128) - and for any denotation d for G, we find an acceptable valuation for G and d.

  The proof consists of a mixed induction, successively assigning values for the X, and splitting up the graph into ever smaller independent subgraphs. The independence of the subgraphs relies essentially on the fact that G (and thus also all subgraphs of G) is simply connected.

(4) In Section 7.3.1 (page 131), we discuss some modifications and generalizations of the Yablo structure. Example 7.3.1 (page 131) considers trivial modifications of the Yablo structure. In Fact 7.3.1 (page 131) (see also Fact 7.1.2 (page 121)) we show that infinite branching is necessary for a graph being dangerous, and Example 7.3.7 (page 143) shows why infinitely many finitely branching points cannot replace infinite branching - there is an infinite "procrastination branch".

  Our main result here is in Fact 7.3.2 (page 132), where we show that in Yablo-like structures, the existence of an acceptable valuation is strongly related to existence of successor nodes, where X' is a successor of X in a directed graph G, iff $X \to X'$ in G, or, written differently, $XX' \in E(G)$, the set of edges in G.

Definition 7.2.1

(1) Call a denotation d $\bigwedge \neg$ or $\bigwedge -$ iff all $d(X)$ have the form $d(X) = \bigwedge\{\neg X_i : i \in I\}$ - as in the Yablo structure.

(2) The dual notation $\bigwedge +$ expresses the analogous case with $+$ instead of $\neg$, i.e. $d(X) = \bigwedge\{X_i : i \in I\}$.

(3) We will use $\neg$ and - for negation, and $+$ when we want to emphasize that a formula is not negated.

Remark 7.2.1

Note that we interpret $\bigwedge$ in the strict sense of $\forall$, i.e., $\neg \bigwedge\{\neg X_i : i \in I\}$ means that there is at least one X_i which is true. In particular, if $d(X) = \bigwedge\{\neg X_i : i \in I\}$, and $[X] = [d(X)] = \bot$, then $d(X)$ must contain a propositional variable, i.e. cannot be composed only of $\bot$ and $\top$, so there is some arrow $X \to X'$ in the graph.

Thus, if in the corresponding graph $succ(X) = \emptyset$, d is of the form $\bigwedge -$, v is an acceptable valuation for d, then $[X]_v = \top$.

Dually, for $\bigwedge +$, $\neg \bigwedge\{X_i : i \in I\}$ means that there is at least one X_i which is false.

Thus, if in the corresponding graph $succ(X) = \emptyset$, d is of the form $\bigwedge +$, v is an acceptable valuation for d, then $[X]_v = \bot$.

7.2.2   A Comment on Conjecture 15 in [RRM13]

We show in this section that conjecture 15 in [RRM13] is wrong.

Definition 7.2.2

Call $\mathcal{X} \subseteq \mathbf{Z}$ (the integers) contiguous iff for all $x, y, z \in \mathbf{Z}$, if $x < y < z$ and $x, z \in \mathcal{X}$, then $y \in \mathcal{X}$, too.

Fact 7.2.2

Let G be a directed graph, $V(G) = \mathcal{X}$ for some contiguous $\mathcal{X}$, and $x_i x_j \in E(G)$ iff x_j is the direct successor of x_i.

Then for any denotation d:

(1) $d(x)$ may be (equivalent to) $x + 1$, $\neg(x + 1)$, $\bot$, or $\top$.
    If $d(x) = \bot$ or $\top$, we abbreviate $d(x) = c$, c', etc. (c for constant).

If v is acceptable for d, then:

(1) If $d(x) = c$, then $d(x - 1) = c'$ (if $x - 1$ exists in $\mathcal{X}$).

(2) if $d(x) = (x + 1)$, then $[x]_v = [x + 1]_v$
    if $d(x) = \neg(x + 1)$, then $[x]_v = \neg[x + 1]_v$

(3) Thus:

(3.1) If $d(x) = c$ for some x, then for all $x' < x$ $d(x') = c'$ for some c'.

(3.2) We have three possible cases:

(3.2.1) $d(x) = c$ for all $x \in \mathcal{X}$,

(3.2.2) $d(x) = c$ for no $x \in \mathcal{X}$,

(3.2.3) there is some maximal x' s.t. $d(x') = c$, so $d(x'') \neq c'$ for all $x'' > x'$.

- In the first case, for all x, if $d(x)$ is $\bot$ or $\top$, then the valuation for x starts anew, i.e. independent of $x + 1$, and continues to $x - 1$ etc. according to (2).

- in the second case, there is just one acceptable valuation: we chose some $x \in \mathcal{X}$, and $[x]_v$ and propagate the value up and down according to (2)

- in the third case, we work as in the first case up to x', and treat the $x'' > x'$ as in the second case.

- Basically, we work downwards from constants, and up and down beyond the maximal constant. Constants interrupt the upward movement.

(4) Consequently, any d on $\mathcal{X}$ has an acceptable valuation v_d, and the graph is not dangerous.

(The present fact is a special case of Fact 7.2.5 (page 128), but it seems useful to discuss a simple case first.)

Example 7.2.1

We define now a modified Yablo graph YG', and a corresponding denotation d, which is paradoxical.

We refer to Fig.3 in [RRM13], and Diagram 7.2.1 (page 127).

(1) The vertices (and the set S of language symbols):

We keep all Y_i of Fig.3 in [RRM13], and introduce new vertices (Y_i, Y_j, Y_k) for $i < k < j$. (When we write (Y_i, Y_j, Y_k), we tacitly assume that $i < k < j$.)

(2) The arrows:

All $Y_i \rightarrow Y_{i+1}$ as before. We "factorize" longer arrows through new vertices:

(2.1) $Y_i \rightarrow (Y_i, Y_j, Y_{i+1})$

(2.2) $(Y_i, Y_j, Y_k) \rightarrow (Y_i, Y_j, Y_{k+1})$

(2.3) $(Y_i, Y_j, Y_{j-1}) \rightarrow Y_j$

See Diagram 7.2.1 (page 127).

We define d (instead of writing $d((x, y, z))$ we write $d(x, y, z)$ - likewise $[x, y, z]_v$ for $[(x, y, z)]_v$ below):

(1) $d(Y_i) := \neg Y_{i+1} \wedge \bigwedge \{\neg(Y_i, Y_j, Y_{i+1}) : i+2 \le j\}$

(This is the main idea of the Yablo construction.)

(2) $d(Y_i, Y_j, Y_k) := (Y_i, Y_j, Y_{k+1})$ for $i < k < j-1$

(3) $d(Y_i, Y_j, Y_{j-1}) := Y_j$

Obviously, YG' corresponds to S and d, i.e. $YG' = G_{S,d}$.

Fact 7.2.3

YG' and d code the Yablo Paradox:

Proof

Let v be an acceptable valuation relative to d.

Suppose $[Y_1]_v = \top$, then $[Y_2]_v = \bot$, and $[Y_1, Y_k, Y_2]_v = \bot$ for $2 < k$, so $[Y_k]_v = \bot$ for $2 < k$, as in Fact 7.2.2 (page 124), (2). By $[Y_2]_v = \bot$, there must be j such that $j = 3$ and $[Y_3]_v = \top$, or $j > 3$ and $[Y_2, Y_j, Y_3]_v = \top$, and as in Fact 7.2.2 (page 124), (2) again, $[Y_j]_v = \top$, a contradiction.

If $[Y_1]_v = \bot$, then as above for $[Y_2]_v$, we find $j \ge 2$ and $[Y_j]_v = \top$, and argue with Y_j as above for Y_1.

Thus, YG' with d as above is paradoxical, and YG' is dangerous.

$\square$

Example 7.2.2

We first define YG'': $V(YG'') := \{\langle Y_i \rangle : i < \omega\}$, $E(YG'') := \{\langle Y_i \rangle \rightarrow \langle Y_{i+1} \rangle : i < \omega\}$.

We now define the homomorphism from YG' to YG''. We collaps for fixed k Y_k and all (Y_i, Y_j, Y_k) to $\langle Y_k \rangle$, more precisely, define f by $f(Y_k) := f(Y_i, Y_j, Y_k) := \langle Y_k \rangle$ for all suitable i, j.

Note that YG' only had arrows between "successor levels", and we have now only arrows from $\langle Y_k \rangle$ to $\langle Y_{k+1} \rangle$, so f is a homomorphism, moreover, our structure YG'' has the form described in Fact 7.2.2 (page 124), and is not dangerous, contradicting conjecture 15 in [RRM13].

Diagram YG'

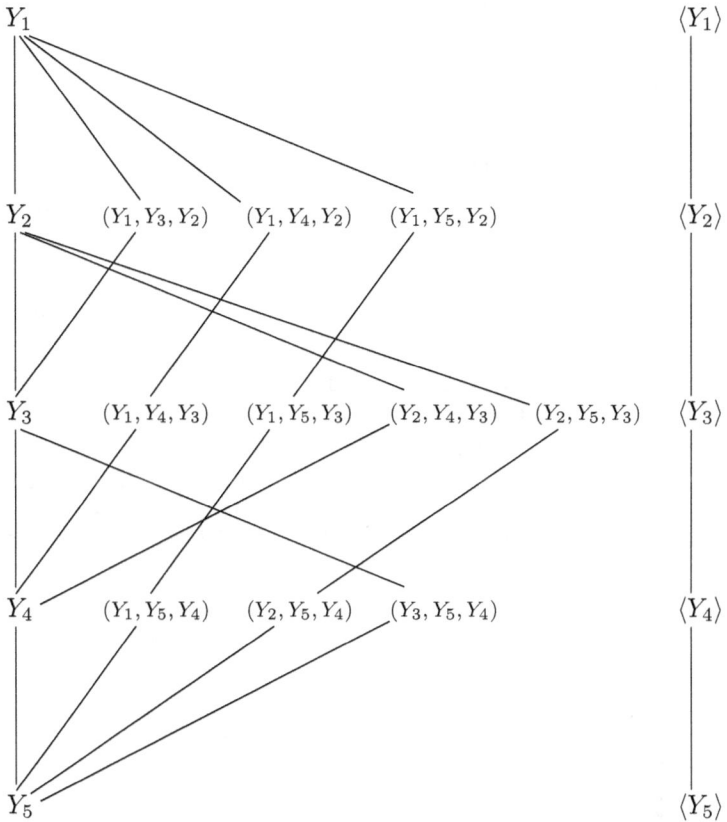

Diagram 7.2.1

This is just the start of the graph, it continues downward through ω many levels.

The lines stand for downward pointing arrows. The lines originating from the Y_i correspond to the negative lines in the original Yablo graph, all others are simple positive lines, of the type $d(X) = X'$.

The left part of the drawing represents the graph YG', the right hand part the collapsed graph, the homomorphic image YG".

Compare to Fig.3 in [RRM13].

7.2.3   A Comment on Theorem 24 of [RRM13]

We comment in this section on the meaning of theorem 24 in [RRM13].

Definition 7.2.3

Fix a denotation d.

Let $s(X) := s(d(X))$ be the set of $s \in S$ which occur in $d(X)$.

Let $r(X) \subseteq s(X)$ be the set of relevant s, i.e. which influence $[d(X)]_v$ for some v.
E.g., in $(\alpha \vee \neg\alpha) \wedge \alpha'$, α' is relevant, α is not.

Definition 7.2.4

(1) Let G be a directed graph. For $X \in G$, let the subgraph $C(X)$ of G be
the connected component of G which contains $X : X \in V(C(X))$, and $X' \in V(C(X))$ iff there is a path in $U(G)$ from X to X', together with the induced
edges of G, i.e., if $Y, Y' \in V(C(X))$, and $YY' \in E(G)$, then $YY' \in E(C(X))$.

(2) G is called a simply connected graph iff for all X, Y in G, there is at most
one path in $U(G)$ from X to Y.

(One may debate if a loop $X \to X$ violates simple connectedness, as we have
the paths $X \to X$ and $X \to X \to X$ - we think so. Otherwise, we exclude
loops.)

(3) Two subgraphs G', G'' of G are disconnected iff there is no path from any
$X' \in G'$ to any $X'' \in G''$ in $U(G)$.

Fact 7.2.4

Let G, d be given, $G = G_{S,d}$.

If G', G'' are two disconnected subgraphs of G, then they can be given truth values
independently.

Proof

Trivial, as the subgraphs share no propositional variables. $\square$

Fact 7.2.5

Let G be simply connected, and d any denotation, $G = G_{S,d}$. Then G, d has an
acceptable valuation.

Proof

This procedure assigns an acceptable valuation to G and d in several steps.

More precisely, it is an inductive procedure, defining v for more and more elements, and cutting up the graph into diconnected subgraphs. If necessary, we will use unions for the definition of v, and the common refinement for the subgraphs in the limit step.

The first step is a local step, it tries to simplify $d(X)$ by looking locally at it, propagating $[X]$ to X' with $X' \to X$ if possible, and erasing arrows from and to X, if possible. Erasing arrows decomposes the graph into disconnected subgraphs, as the graph is simply connected.

The second step initializes an arbitrary value X (or, in step (4), uses a value determined in step (2)), propagates the value to X' for $X' \to X$, erases the arrow $X' \to X$. Initialising X will have repercussions on the X'' for $X \to X''$, so we chose a correct possibility for the X'' (e.g., if $d(X) = X'' \wedge X'''$, setting $[X] = \top$, requires to set $[X''] = [X'''] = \top$, too), and erase the arrows $X \to X''$. As G is simply connected, the only connection between the different $C(X'')$ is via X, but this was respected and erased, and they are now independent.

(1) Local step

    (1.1) For all $X' \in s(X) - r(X)$:

        (1.1.1) replace X' in $d(X)$ by $\top$ (or, equivalently, $\bot$), resulting in logically equivalent $d'(X)$ ($s(X')$ might now be empty),

        (1.1.2) erase the arrow $X \to X'$.
            Note that $C(X')$ will then be disconnected from $C(X)$, as G is simply connected.

    (1.2) Do recursively:
        If $s(d(X)) = \emptyset$, then $d(X)$ is equivalent to $\top$ (or $\bot$) (it might also be $\top \wedge \bot$ etc.), so $[X]_v = [d(X)]_v = \top$ (or $\bot$) in any acceptable valuation, and $[d(X)]_v$ is independent of v.

        (1.2.1) For $X' \to X$, replace X in $d(X')$ by $\top$ (or $\bot$) ($s(X')$ might now be empty),

        (1.2.2) erase $X' \to X$ in G.
            X is then an isolated point in G, so its truth value is independent of the other truth values (and determined already).

(2) Let G'' be a non-trivial (i.e. not an isolated point) connected component of the original graph G, chose X in G''. If X were already fixed as $\top$ or $\bot$, then X would have been isolated by step (1). So $[X]_v$ is undetermined so far. Moreover, if $X \to X'$ in G'', then $d(X')$ cannot be equivalent to a constant value either, otherwise, the arrow $X \to X'$ would have been eliminated already in step (1).

Chose arbitrarily a truth value for $d(X)$, say $\top$.

    (2.1) Consider any X' s.t. $X' \to X$ (if this exists)

        (2.1.1) Replace X in $d(X')$ with that truth value, here $\top$.

(2.1.2) Erase $X' \to X$

As G'' is simply connected, all such $C(X')$ and $C(X)$ are now mutually disconnected.

(2.2) Consider simultanously all X'' s.t. $X \to X''$. (They are not constants, as any $X'' \in V(G'')$ must be a propositional variable.)

(2.2.1) Chose values for all such X'', corresponding to $[X]_v = [d(X)]_v \ (= \top$ here).

E.g., if $d(X) = X'' \wedge X'''$, and the value for X was $\top$, then we have to chose $\top$ also for X'' and X'''.

This is possible independently by Fact 7.2.4 (page 128), as the graph G'' is simply connected, and X is the only connection between the different X''

(2.2.2) Erase all such $X \to X''$.

X is now an isolated point, and as G'' is simply connected, all $C(X'')$ are mutually disconnected, and disconnected from all $C(X')$ with $X' \to X$, considered in (2.1).

The main argument here is that we may define $[X'']_v$ and $[X''']_v$ for all $X \to X''$ and $X \to X'''$ independently, if we respect the dependencies resulting through X.

(3) Repeat step (1) recursively on all mutually disconnected fragments resulting from step (2).

(4) Repeat step (2) for all X'' in (2.2), but instead of the free choice for $[X]_v$ in (2), the choice for the X'' has already been made in step (2.2.1), and work with this choice.

$\square$

7.3   Remarks on Contradictory Structures

7.3.1   Introductory Comments

Example 7.3.1

We discuss here some very simple examples, all modifications of the Yablo structure.

Up to now, we considered graphs isomorphic to (parts of) the natural numbers with arrows pointing to bigger numbers. We consider now other cases, and describe the underlying graphs. Arrows are understood as negative.

(1) Consider the negative numbers (with 0), arrows pointing again to bigger numbers. Putting + at 0, and - to all other elements is an acceptable valuation.

(2) Consider a tree with arrows pointing to the root. The tree may be infinite. Again + at the root, - at all other elements is an acceptable valuation.

(3) Consider an infinite tree, the root with ω successors x_i, $i < \omega$, and from each x_i originating a chain of length i as in Fig. 10 of [RRM13], putting + at the end of the branches, and - everywhere else is an acceptable valuation.

(4) This trivial example shows that an initial segment of a Yablo construction can again be a Yablo construction.

Instead of considering all Y_i, $i < \omega$, we consider Y_i, $i < \omega + \omega$, extending the original construction in the obvious way.

Fact 7.3.1

Let G be loop free and finitely forward branching, i.e. for any s, there are only finitely many s' such that $s \to s'$ in G. Then G is not dangerous.

(d may be arbitrary, not necessarily of the $\bigwedge -$ form, i.e. $\bigwedge \neg x_i$.)

See also Fact 7.1.2 (page 121).

Proof

Let d be any assignment corresponding to G. Then $d(s)$ is a finite, classical formula. Replace $[s]_v = [d(s)]_v$ by the classical formula $\phi_s := s \leftrightarrow d(s)$. Then any finite number of ϕ_s is consistent.

Proof: Let Φ be a finite set of such ϕ_s, and S_Φ the set of s occurring in Φ. As G is loop free, and S_Φ finite, we may initialise the minimal $s \in S_\Phi$ (i.e. there is no s' such that $s \to s'$ in the part of G corresponding to Φ) with any truth values, and propagate the truth values upward according to usual valuation rules. This shows that Φ is consistent, i.e. we have constructed a (partial) acceptable valuation for d.

Extend Φ by classical compactness, resulting in a total acceptable valuation for d.

(In general, in the logics considered here, compactness obviously does not hold: Consider $\{\neg \bigwedge \{Y_i : i < \omega\} \cup \{Y_i : i < \omega\}$. Clearly, every finite subset is consistent, but the entire set is not.)

$\square$

Fact 7.3.2

Let G be transitive, and d be of the type $\bigwedge -$.

(1) If $\exists X. \ (succ(X) \neq \emptyset$ and $\forall X' \in succ(X).succ(X') \neq \emptyset)$, then d has no acceptable valuation.

Let acceptable v be given, $[.]$ is for this v.

Case 1: $[X] = \top$. So for all $X' \in succ(X)$ $[X'] = \bot$, and there is such X', so (either by the prerequisite $succ(X') \neq \emptyset$, or by Remark 7.2.1 (page 124)) $\exists$ $X'' \in succ(X').[X''] = \top$, but $succ(X') \subseteq succ(X)$, a contradiction.

In abbreviation: $X^+ \rightarrow_{\bigwedge -} X'^- \rightarrow_{\bigwedge -} X''^+$

Case 2: $[X] = \bot$. So $\exists X' \in succ(X).[X'] = \top$, so $\forall X'' \in succ(X').[X''] = \bot$, and by prerequisite $succ(X') \neq \emptyset$, so there is such X'', so by Remark 7.2.1 (page 124) $succ(X'') \neq \emptyset$, so $\exists\ X''' \in succ(X'').[X'''] = \top$, but $succ(X'') \subseteq succ(X')$, a contradiction.

$X^- \rightarrow_{\bigwedge -} X'^+ \rightarrow_{\bigwedge -} X''^- \rightarrow_{\bigwedge -} X'''^+$

(Here we need Remark 7.2.1 (page 124) for the additional step from X'' to X'''.)

(2) Conversely:

Let $\forall X \ (succ(X) = \emptyset$ or $\exists X' \in succ(X).succ(X') = \emptyset)$:

By Remark 7.2.1 (page 124), if $succ(Y) = \emptyset$, then for any acceptable valuation, $[Y] = \top$. Thus, if there is $X' \in succ(X)$, $succ(X') = \emptyset$, $[X'] = \top$, and $[X] = \bot$.

Thus, the valuation defined by $[X] = \top$ iff $succ(X) = \emptyset$, and $\bot$ otherwise is an acceptable valuation. (Obviously, this definition is free from contradictions.)

$\square$

7.3.2   Basics For a more Systematic Investigation

7.3.2.1   Some Trivialities

We will work here with disjunctive normal forms, i.e. with formulas of the type $a := \bigvee \{\bigwedge a_i : i \in I\}$, where $a_i := \{a_{i,j} : j \in J_i\}$, and the $a_{i,j}$ are propositional variables or negations thereof.

Fact 7.3.3

Let $a := \bigvee \{\bigwedge a_i : i \in I\}$, where $a_i := \{a_{i,j} : j \in J_i\}$, and the $a_{i,j}$ are propositional variables or negations thereof.

(1) Let $F := \Pi\{a_i : i \in I\}$.

Then $\neg a = \bigvee \{\bigwedge \{\neg a_{i,j} : a_{i,j} \in ran(f)\} : f \in F\}$.

(By the laws of distributivity.)

(2) Contradictions will be between two formulas only, one a propositional variable, the other the negation of the former.

□

For illustration, we develop Example 7.3.2 (page 133), to see how this works. This particularly useful as an ilustration for Section 7.4.2 (page 155).

Example 7.3.2

Consider the following situation:

$x = (a \wedge b) \vee (c \wedge d)$

$a = (aa \wedge ab) \vee (ac \wedge ad)$

$b = (ba \wedge bb) \vee (bc \wedge bd)$

$c = (ca \wedge cb) \vee (cc \wedge cd)$

$d = (da \wedge db) \vee (dc \wedge dd)$

(1) Conjunction

(1.1) $a \wedge b = [(aa \wedge ab) \vee (ac \wedge ad)] \wedge [(ba \wedge bb) \vee (bc \wedge bd)] =$
$[(aa \wedge ab) \wedge (ba \wedge bb)] \vee [(aa \wedge ab) \wedge (bc \wedge bd)] \vee [(ac \wedge ad) \wedge (ba \wedge bb)] \vee [(ac \wedge ad) \wedge (bc \wedge bd)]$

a has 2 components, $a_1 = (aa \wedge ab)$ and $a_2 = (ac \wedge ad)$, analogously $b_1 = (ba \wedge bb)$ and $b_2 = (bc \wedge bd)$.

Distributivity results in choice functions in the components: $(a_1 \wedge b_1) \vee (a_1 \wedge b_2) \vee (a_2 \wedge b_1) \vee (a_2 \wedge b_2)$.

(1.2) $c \wedge d = [(ca \wedge cb) \vee (cc \wedge cd)] \wedge [(da \wedge db) \vee (dc \wedge dd)] =$
$[(ca \wedge cb) \wedge (da \wedge db)] \vee [(ca \wedge cb) \wedge (dc \wedge dd)] \vee [(cc \wedge cd) \wedge (da \wedge db)] \vee [(cc \wedge cd) \wedge (dc \wedge dd)]$

(2) Negation

(2.1) $\neg x = \neg((a \wedge b) \vee (c \wedge d)) = \neg(a \wedge b) \wedge \neg(c \wedge d) = (\neg a \vee \neg b) \wedge (\neg c \vee \neg d)$
$=$
$(\neg a \wedge \neg c) \vee (\neg a \wedge \neg d) \vee (\neg b \wedge \neg c) \vee (\neg b \wedge \neg d)$

Negation works with distributivity, thus with choice functions, here in the 2 components $x_1 = a \wedge b$, $x_2 = c \wedge d$.

The elements of the components stay the same, only the sign changes.

(2.2) $\neg a = \neg((aa \wedge ab) \vee (ac \wedge ad)) = \neg(aa \wedge ab) \wedge \neg(ac \wedge ad) = (\neg aa \vee \neg ab)$
$\wedge (\neg ac \vee \neg ad) =$
$(\neg aa \wedge \neg ac) \vee (\neg aa \wedge \neg ad) \vee (\neg ab \wedge \neg ac) \vee (\neg ab \wedge \neg ad)$

(2.3) $\neg b = \neg((ba \wedge bb) \vee (bc \wedge bd)) = \neg(ba \wedge bb) \wedge \neg(bc \wedge bd) = (\neg ba \vee \neg bb) \wedge$
$(\neg bc \vee \neg bd) =$
$(\neg ba \wedge \neg bc) \vee (\neg ba \wedge \neg bd) \vee (\neg bb \wedge \neg bc) \vee (\neg bb \wedge \neg bd)$

(2.4) $\neg c = \neg((ca \wedge cb) \vee (cc \wedge cd)) = \neg(ca \wedge cb) \wedge \neg(cc \wedge cd) = (\neg ca \vee \neg cb) \wedge$
$(\neg cc \vee \neg cd) =$
$(\neg ca \wedge \neg cc) \vee (\neg ca \wedge \neg cd) \vee (\neg cb \wedge \neg cc) \vee (\neg cb \wedge \neg cd)$

(2.5) $\neg d = \neg((da \wedge db) \vee (dc \wedge dd)) = \neg(da \wedge db) \wedge \neg(dc \wedge dd) = (\neg da \vee \neg db)$
$\wedge (\neg dc \vee \neg dd) =$
$(\neg da \wedge \neg dc) \vee (\neg da \wedge \neg dd) \vee (\neg db \wedge \neg dc) \vee (\neg db \wedge \neg dd)$

(3) Conjunction of negations

$\neg a \wedge \neg c =$

$[(\neg aa \wedge \neg ac) \vee (\neg aa \wedge \neg ad) \vee (\neg ab \wedge \neg ac) \vee (\neg ab \wedge \neg ad)] \wedge$
$[(\neg ca \wedge \neg cc) \vee (\neg ca \wedge \neg cd) \vee (\neg cb \wedge \neg cc) \vee (\neg cb \wedge \neg cd)] =$

$[(\neg aa \wedge \neg ac) \wedge (\neg ca \wedge \neg cc)] \vee [(\neg aa \wedge \neg ac) \wedge (\neg ca \wedge \neg cd)] \vee$
$[(\neg aa \wedge \neg ac) \wedge (\neg cb \wedge \neg cc)] \vee [(\neg aa \wedge \neg ac) \wedge (\neg cb \wedge \neg cd)] \vee$
$[(\neg aa \wedge \neg ad) \wedge (\neg ca \wedge \neg cc)] \vee [(\neg aa \wedge \neg ad) \wedge (\neg ca \wedge \neg cd)] \vee$
$[(\neg aa \wedge \neg ad) \wedge (\neg cb \wedge \neg cc)] \vee [(\neg aa \wedge \neg ad) \wedge (\neg cb \wedge \neg cd)] \vee$
$[(\neg ab \wedge \neg ac) \wedge (\neg ca \wedge \neg cc)] \vee [(\neg ab \wedge \neg ac) \wedge (\neg ca \wedge \neg cd)] \vee$
$[(\neg ab \wedge \neg ac) \wedge (\neg cb \wedge \neg cc)] \vee [(\neg ab \wedge \neg ac) \wedge (\neg cb \wedge \neg cd)] \vee$
$[(\neg ab \wedge \neg ad) \wedge (\neg ca \wedge \neg cc)] \vee [(\neg ab \wedge \neg ad) \wedge (\neg ca \wedge \neg cd)] \vee$
$[(\neg ab \wedge \neg ad) \wedge (\neg cb \wedge \neg cc)] \vee [(\neg ab \wedge \neg ad) \wedge (\neg cb \wedge \neg cd)]$

$\neg a$ and $\neg c$ have 4 components each, we chose in the 4 components.

7.3.2.2   Cells and Contradictions

Definition 7.3.1

(1) A cell is a set of labelled paths with the following properties:

(1.1) they have a common origin (this is not essential, it simplifies the definition slightly).

(1.2) for each pair of paths there is a point x, where they diverge,

(1.3) for each pair of paths, once they diverged, they meet again, but will not diverge again.

(2) A contradictory cell is a cell such that there is at least one valuation and one pair of paths whose valued versions contradict each other.

Example 7.3.3

(1) The following are cells:

(1.1) $x \to y,\ x \not\to y$

(1.2) $x \to y \to z,\ x \to z$

(1.3) $x \not\to y \not\to z,\ x \not\to z$

(1.4) $x \not\to y \not\to z \to w,\ x \not\to z \to w$

(1.5) $w \to x \not\to y \not\to z,\ w \to x \not\to z$

(1.6) $u \to x \not\to x' \to x'' \to z,\ u \to x \to x'' \to z,\ u \to y \not\to y' \to y'' \to z,$
$u \to y \to y'' \to z$

Note that this cell is composed of two sub-cells, which are in parallel.

(2) The following are not cells, as they diverge again:

(2.1) $x \to y \to z,\ x \not\to y \not\to z$

(2.2) $x \not\to y \not\to z \not\to y' \not\to z',\ z \not\to z \not\to z'$

These cells are serially connected.

(3) In Example 7.4.1 (page 154), the following are cells:

(3.1) $x_0 \not\to x_1 \to x_2,\ x_0 \to x_2,$

(3.2) $x_0 \not\to x_1 \to x_2 \to x_3,\ x_0 \to x_2 \to x_3,$

(3.3) $x_0 \not\to x_1 \to x_2 \to x_3 \to x_4,\ x_0 \to x_2 \to x_3 \to x_4,\ x_0 \to x_4$

(3.4) $x_2 \to x_3,\ x_2 \not\to x_3$

(3.5) $x_0 \to x_2 \to x_3,\ x_0 \to x_2 \not\to x_3$

(3.6) and the following is not a cell
$x_0 \not\to x_1 \to x_2 \not\to x_3,\ x_0 \to x_2 \to x_3$
(after meeting at x_2, the paths continue differently).

Fact 7.3.4

Let σ and τ be two paths with same origin, which meet again at some x and are contradictory. If $v(\sigma_0) = v(\tau_0)$, and $v(\sigma_x) \neq v(\tau_x)$, then the number of negative arrows in σ between σ_0 and σ_x modulo 2 is unequal to the number of negative arrows in τ between τ_0 and τ_x modulo 2.

The following example is very important, and the basis for the original Yablo construction, as well as our "saw blade" construction.

Example 7.3.4

See also Example 7.6.2 (page 180).

We consider now some simple, contradictory cells. They should not only be contradictory for the case $x+$, but also be a potential start for the case $x-$, without using more complex cells.

For this, we order the complexity of the cases by $(1) < (2.1) < (2.3)$ below, (2.2) is not contradictory, so it is excluded.

See Diagram 7.3.2 (page 139).

(1) The cell with 2 arrows.

It corresponds to the formula $d(x) = y \wedge \neg y$, graphically, it has a positive and a negative arrow from x to y, so exactly one of α and β is negative.

If x is positive, we have a contradiction.

If x is negative, however, we have a problem. Then, we have $d(x) = y \vee \neg y$. Let α be the originally positive path, β the originally negative path. Note that α is now negative, and β is positive. The α presents no problem, as y is positive, and we can append the same construction to y, and have a contradiction. However, β has to lead to a contradiction, too, and, as we will not use more complicated cells, we face the same problem again, y is negative. So we have an "escape path", assigning $\bot$ to all elements in one branch.

(Consider $x_0 \Rightarrow_\neg x_1 \Rightarrow_\neg x_2 \Rightarrow_\neg x_3 \ldots$, setting all $x_i := -$ is a consistent valuation. So combining this cell with itself does not result in a contradictory structure.)

Of course, appending at y a Yablo Cell (see below, case (2.3)) may be the beginning of a contradictory structure, but this is "cheating", we use a more complex cell.

(2) Cells with 3 arrows.

Note that the following examples are not distinguished in the graph!

Again, we want a contradiction for x positive, so we need an $\wedge$ at x, $d(x)$ (or x, we abbreviate) and have the possibilities (up to equivalence) $x = x' \wedge y$, $x = x' \wedge \neg y$, $x = \neg x' \wedge y$, $x = \neg x' \wedge \neg y$.

Again, if x is negative, all paths $\alpha : x \to y$, $\beta : x \to x'$ (or $\beta \gamma : x \to x' \to y$) have to lead to a contradiction.

(2.1) one negative arrow, with $d(x) = \pm x' \wedge \pm y$

(2.1.1) $x \to x' \nrightarrow y$, $x \to y$, corresponding to $x = x' \wedge y$, so if x is negative, this is $\neg x' \vee \neg y$.

But, at y, this possible paths ends, and we have the same situation again, with a negative start, as in Case (1).

(2.1.2) $x \to x' \to y, x \not\to y$

   Here, we have again a positive path to y, through x', so both x' and y will be negative, and neither gives a start for a new contradiction.

(2.1.3) $x \not\to x' \to y, x \to y$

   This case is analogous to case (2.1.1).

   (Similar arguments apply to more complicated cells with an even number of negative arrows until the first branching point - see Remark 7.3.5 (page 140) below and the discussion of the Yablo construction, Section 7.3.2.3 (page 142).

(2.2) 2 negative arrows: not contradictory.

(2.3) The original type of contradiction in Yablo's construction

   $x \not\to x' \not\to y, x \not\to y$.

   This will be discussed in detail below in Section 7.3.2.3 (page 142). But we see already that both paths, $x \not\to x'$ and $x \not\to y$ change sign, so x' and y will be positive, appending the same type of cell at x' and y solves the problem (locally), and offers no escape.

Definition 7.3.2

(1) We call the contradiction of the type (2.3) of Example 7.3.4 (page 136), i.e. $x \not\to x' \not\to y, x \not\to y$, a Yablo Cell, YC, and sometimes x its head, y its foot, and x' its knee.

(2) We sometimes abbreviate a Yablo Cells simply by ∇, without going into any further details.

If we combine Yablo Cells, the knee for one cell may become the head for another Cell, etc.

See Diagram 7.3.1 (page 138).

Yablo Cells

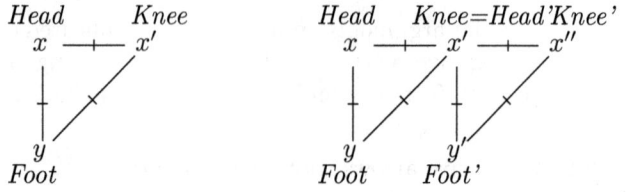

Diagram 7.3.1

Contradictory Cells

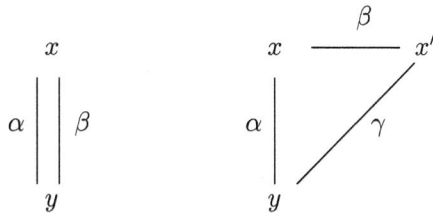

Diagram 7.3.2

See Remark 7.3.5

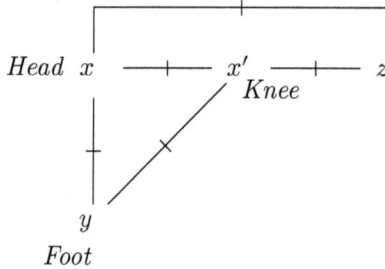

Diagram 7.3.3

Remark 7.3.5

(1) The distinction between x' and y, i.e. between knee and foot, is very important. In the case $x+$, we have at y a complete contradiction, at x', we have
not yet constructed a contradiction. Thus, if we have at x' again an $\wedge$ (as at
x), this becomes an $\vee$, and we have to construct a contradiction for all $x' \to z$
(or $x' \not\to z$), not only for $x' \not\to y$. Otherwise, we have an escape possibility for
$x+$. Obviously, the contradiction need not be immediate at z, the important
property is that ALL paths through all z lead to a contradiction, and the
simplest way is to have the contradiction immediately at z - as in Yablo's
construction, and our saw blades.

See Diagram 7.3.3 (page 140).

See also the discussion in Fact 7.6.3 (page 173) and the construction of saw
blades in Section 7.6.2 (page 168).

(2) As we work for a contradiction in the case x-, too, the simplest way to achieve
this is to have a negative arrow $x \not\to x'$, and at x' again an $\wedge$. This gives a
chance to construct a contradiction at x'. Of course, we have to construct a
contradiction at y, too, as in the case $x-$, we have an $\vee$ at x.

Of course, we may have branches originating at x', which all lead to contradictions in the case $x-$, so $x \to x'$ (resp. $\vee$ at x') is possible, too.

But, for the simple construction, we need $\not\to$ between x and x', and $\wedge$ at x'. And this leads to the construction of contradictions for all $x' \to z$ (or $x' \not\to z$) as just mentioned above.

7.3.2.3   Further Comments On the Yablo Construction and Variations

We start with a Yablo Cell (YC) $x_0 \nrightarrow x_1 \nrightarrow x_2$, $x_0 \nrightarrow x_2$.

Note that in both cases, initialising x_0 with + or with -, the paths $x_0 \nrightarrow x_1 \nrightarrow x_2$, $x_0 \nrightarrow x_2$ are contradictory, but in the case +, with $\wedge$ at x_0, we "feel" the contradiction, in the case - with $\vee$ at x_0, the contradiction is irrelevant.

We are mainly interested in solutions which may be repeated without any modification.

(1) If x_0+, then x_1- and $\vee$. Thus, to have a contradiction at x_0+, any y with $x_1 \nrightarrow y$ (or $x_1 \rightarrow y$) has to be contradictory, either directly at x_0 (as is the case for $x_1 \nrightarrow x_2$), or in some other way, for say x_0'.

If we repeat the latter construction with x_0', then we may construct an escape path $x_0 \ldots x_0' \ldots x_0'' \ldots$ for the $\vee$'s, i.e. an infinite choice for the $\vee$'s, never leading to a contradiction.

Example 7.3.5

(1.1) $x_0 \nrightarrow x_1 \nrightarrow x_1' \nrightarrow x_3 \nrightarrow x_3' \ldots$, $x_0 \nrightarrow x_2$, $x_1 \nrightarrow x_2$, $x_1' \nrightarrow x_4$, $x_3 \nrightarrow x_4$ etc. with $x_0 \nrightarrow x_1 \nrightarrow x_1' \nrightarrow x_3 \nrightarrow x_3' \ldots$ the escape route. (The $x_1 \nrightarrow x_1'$ serve to have at x_1' again an $\wedge$).

(1.2) $x_0 \nrightarrow x_1 \nrightarrow x_3 \nrightarrow x_5 \ldots$, $x_0 \nrightarrow x_2$, $x_1 \nrightarrow x_2$, $x_1 \nrightarrow x_4$, $x_3 \nrightarrow x_4$, $x_3 \nrightarrow x_6$, $x_5 \nrightarrow x_6$ etc. with $x_0 \nrightarrow x_1 \nrightarrow x_3 \nrightarrow x_5 \ldots$ the escape route. This is similar to (1), the escape is constructed at x_1, x_3 is an $\wedge$, but we have an escape again at x_5, etc.

- We repair the new possibilities $x_1 \nrightarrow y$ by direct contradictions $x_0 \nrightarrow y$, avoiding procrastination recursively leading to other contradictions.
  (We indroduced a new arrow $x_1 \nrightarrow y$, and have to repair the new possibility for the OR-case in x_1 by the arrow $x_0 \nrightarrow y$.)

(2) If we simplify, we might use the arrow $x_1 \nrightarrow x_2$, instead of starting anew e.g. as above with $x_1 \nrightarrow x_3 \nrightarrow x_4$, $x_1 \nrightarrow x_4$ in case (2) of Example 7.3.5 (page 142).

We have two possibilities:

(2.1) Use $x_1 \nrightarrow x_2$ for the "long" arrow, add $x_1 \nrightarrow x_3 \nrightarrow x_2$.
He This leads to the problem that x_2 becomes a smallest element, and we may insert truth values bottom to top:
$x_0 \nrightarrow x_1 \nrightarrow x_3 \nrightarrow x_4 \nrightarrow x_5 \ldots$, $x_1 \nrightarrow x_2$, $x_3 \nrightarrow x_2$, $x_4 \nrightarrow x_2$, $x_5 \nrightarrow x_2$, etc.
So, this solution is not "sustainable". (Of course, we may continue below x_2 as for x_0. But this is unnecessarily complicated.)

(2.2) Use $x_1 \nrightarrow x_2$ for the "short" arrow, add $x_1 \nrightarrow x_3$, $x_2 \nrightarrow x_3$, etc.
This is the original Yablo construction.

(2.2.1) As the truth value of x_3 must not be definable, x_3 has to be the head of a YC, so there must be $x_3 \not\to x_4 \not\to x_5$, $x_3 \not\to x_5$, each again heads of YC's.

- This forces infinite depth, so we cannot fill values from below.

But as x_2 is the head of a YC, by (1.2) above, we have to add arrows $x_2 \not\to x_4$ and $x_2 \not\to x_5$.

Recursively, this forces us again by (1.2) above to add arrows $x_1 \not\to x_4$ and $x_1 \not\to x_5$, and then to add arrows $x_0 \not\to x_4$ and $x_0 \not\to x_5$.

- This forces infinite branching (width).

See also Example 7.3.6 (page 143).

(2.2.2) Note that it is not necessary to keep the whole construction once created, it is enough to keep suitable fragments cofinally often.

(3) Adding constants:

Suppose we have $x \not\to x'$, $x \not\to x''$, ... and add the arrow $x \not\to TRUE$. Then x becomes FALSE (because of $\not\to$).

Suppose we have $x \not\to x'$, $x \not\to x''$, ... and add the arrow $x \not\to FALSE$. Then x is as if the new arrow does not exist, except when there are no x' etc., then x becomes TRUE (because of $\not\to$).

Of course, this propagates upward (and downward).

(4) We emphasize again the importance of

(4.1) repairing added arrows
and

(4.2) avoiding procrastination, see also Example 7.3.7 (page 143)

Example 7.3.6

$x_0 \not\to x_1 \not\to x_2 \not\to x_3 \not\to x_4$, $x_0 \not\to x_2$, $x_0 \not\to x_3$,

$x_1 \not\to x_3$, $x_1 \not\to x_4$,

$x_2 \not\to x_4$,

$x_3 \not\to x_4$,

add $x_0 \not\to x_4$ in the last step.

The following Example 7.3.7 (page 143) shows that infinitely many finitely branching points cannot always replace infinite branching - there is an infinite "procrastination branch" or "escape branch". This modification of the Yablo structure has one acceptable valuation for Y_1 :

Example 7.3.7

Let Y_i, $i < \omega$ as usual, and introduce new X_i, $3 \leq i < \omega$.

Let $Y_i \to Y_{i+1}$, $Y_i \to X_{i+2}$, $X_i \to Y_i$, $X_i \to X_{i+1}$, with

$d(Y_i) := \neg Y_{i+1} \wedge X_{i+2}$, $d(X_i) := \neg Y_i \wedge X_{i+1}$.

If $Y_1 = \top$, then $\neg Y_2 \wedge X_3$, by X_3, $\neg Y_3 \wedge X_4$, so, generally,

if $Y_i = \top$, then $\{\neg Y_j : i < j\}$ and $\{X_j : i+1 < j\}$.

If $\neg Y_1$, then $Y_2 \vee \neg X_3$, so if $\neg X_3$, $Y_3 \vee \neg X_4$, etc., so, generally,

if $\neg Y_i$, then $\exists j (i < j, Y_j)$ or $\forall j \{\neg X_j : i+1 < j\}$.

Suppose now $Y_1 = \top$, then X_j for all $2 < j$, and $\neg Y_j$ for all $1 < j$. By $\neg Y_2$ there is j, $2 < j$, and Y_j, a contradiction, or $\neg X_j$ for all $3 < j$, again a contradiction.

But $\neg Y_1$ is possible, by setting $\neg Y_i$ and $\neg X_i$ for all i.

Thus, replacing infinite branching by an infinite number of finite branching does not work for the Yablo construction, as we can always chose the "procrastinating" branch.

See Diagram 7.3.4 (page 145).

We can turn this into a trivial little fact:

Fact 7.3.6

If, at each stage, we leave some work undone, we construct an escape path, following which never finish work. (Obvious, or look at the proof of Koenig's Infinity Lemma (in set theory).)

We have to be careful, however. This applies to an ever branching situation, but not to a set of branches given at the outset, where we eliminate one branch after the other. Consider the classic example, where we have ω many branches of height n for branch n, branching directly off the root, which we eliminate one after the other. There is no escape branch here.

□

Diagram for Example 7.3.7

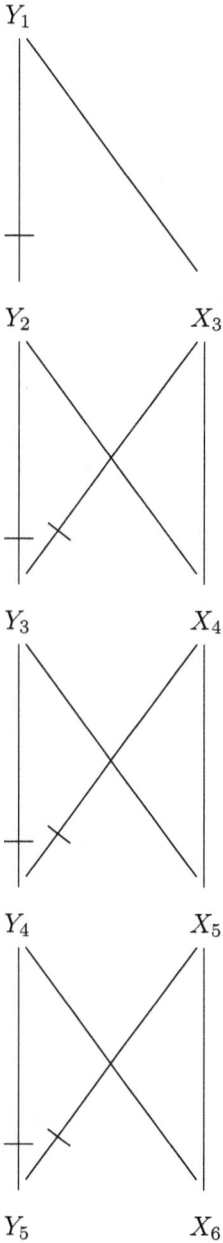

Y_1

Y_2 X_3

Y_3 X_4

Y_4 X_5

Diagram 7.3.4 Y_5 X_6

7.3.2.4   More Complicated Systems of Contradictions

We consider now more complex cells, built up, as before, from negative arrows and conjunctions.

Notation 7.3.1

We may write $x-$, $x\vee$, $x-\vee$, etc., to abbreviate that the truth value of x is negative, we have an $"\vee"$ at x, both hold, etc.

This is useful to find one's way through the different cases and valuations.

We start again from x_0. If x_0-, then $x_0\vee$, and each branch starting at x_0 has to lead directly or indirectly to a contradiction, i.e. a contradiction cell.

We concentrate on x_0+, so we need a contradiction at x_0.

(1) In the simplest case, with no branching except at x_0, we have a contradiction: $x_0 \not\rightarrow x_1 \not\rightarrow x_2$, $x_0 \not\rightarrow x_2$, the Yablo Cell. See Diagram 7.3.5 (page 147).

(2) The following situation is hardly more complicated: $x_0 \not\rightarrow x_1 \not\rightarrow x_2 \not\rightarrow x_3$, $x_0 \not\rightarrow y_1 \not\rightarrow x_3$. See Diagram 7.3.6 (page 148).

(3) In the situation of the Yablo construction, we branch at x_1, but as $x_1 - \vee$, we do not know if x_1- holds because of x_2+, or because of $x_2'+$, so we have two branches originating in x_0, which both have to be contradictory. See Diagram 7.3.7 (page 149).

(4) Suppose we do not branch at x_1, but at x_2, and have no contradiction at x_2 :

$x_0 \not\rightarrow x_1 \not\rightarrow x_2 \not\rightarrow x_3$, $x_2 \not\rightarrow x_3'$, $x_0 \not\rightarrow y_1 \not\rightarrow x_3$. See Diagram 7.3.8 (page 150).

If x_0+, then x_2+ by $x_0 \not\rightarrow x_1 \not\rightarrow x_2$, and x_3+ by $x_0 \not\rightarrow y_1 \not\rightarrow x_3$. But, as $x_2+ = \neg x_3 \wedge \neg x_3'$, we have a contradiction, because $x_2 + \wedge$. So we do not need some $x_0 \not\rightarrow y_1' \not\rightarrow x_3'$.

(5) Our investigation is asymmetrical, as we concentrate on the $x_i's$. For instance, if we were to branch at y_1, then the branch leading not to x_3 would also have to be contradictory, e.g. by:

$x_0 \not\rightarrow x_1 \not\rightarrow x_2 \not\rightarrow x_3$, $x_2 \not\rightarrow x_3'$, $x_0 \not\rightarrow y_1 \not\rightarrow x_3$, $y_1 \not\rightarrow x_3'$. See Diagram 7.3.9 (page 151).

(6) Take this one step further:

$x_0 \not\rightarrow x_1 \not\rightarrow x_2 \not\rightarrow x_3 \not\rightarrow x_4$, $x_3 \not\rightarrow x_4$, $x_3 \not\rightarrow x_4'$, $x_2 \not\rightarrow x_3'$, $x_0 \not\rightarrow x_4$, $x_0 \not\rightarrow x_4'$. See Diagram 7.3.10 (page 152).

$x_0 \not\rightarrow x_4$ alone would not be enough, we have to add $x_0 \not\rightarrow x_4'$, as $x_3 - \vee$, and we do not know which of x_4+ or $x_4'+$ holds.

(7) So we check formulas with alternating quantifiers for contradictions. The author does not know if there are known facts about this question.

C1

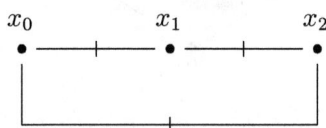

Diagram 7.3.5

C2

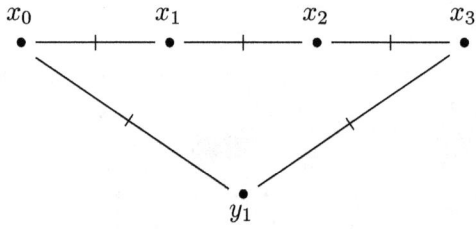

Diagram 7.3.6

C3

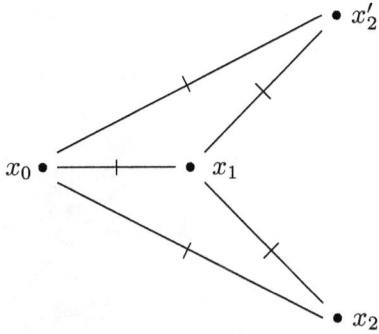

Diagram 7.3.7

C4

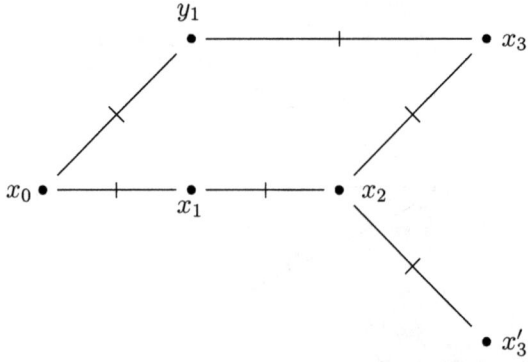

Diagram 7.3.8

C5

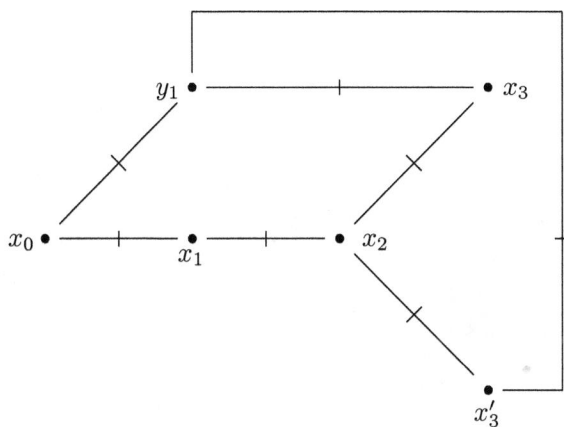

Diagram 7.3.9

C6

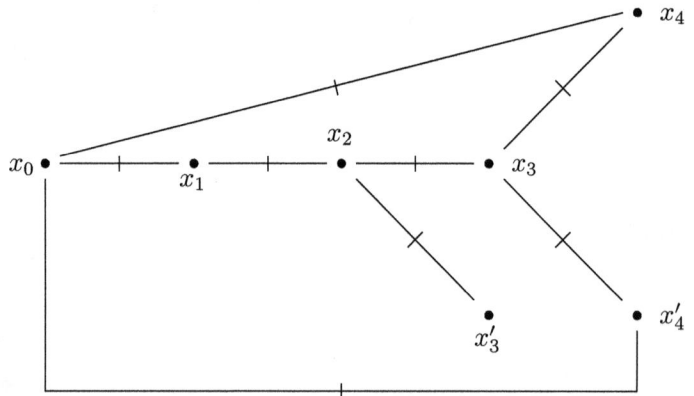

Diagram 7.3.10

7.4   Valuations and Cycles

Usually, it is not necessary to write down the details of valuations, and we may avoid additional notation. Here, we treat somewhat complicated cases, where this notation is useful.

Our main result here is Proposition 7.4.3 (page 158), which illustrates that we cannot achieve impossibility of valuation with finite depth formulas. We know this, but our result is more constructive. We show that any attempt to construct such a set of formulas without consistent valuation will construct a cycle of uneven length of contradictory paths, which is impossible, basically due to the fact that we have 2 truth values.

7.4.1   Paths

Definition 7.4.1

Let $\mathcal{G}$ be a graph as in the Yablo structure.

(1) Label of a node

Call $d(x)$ as in [RRM13] the label of node x.

(2) Labelled arrow

We work with $\bigvee \bigwedge$.

Consider the example $d(x) = (y \wedge z) \vee (\neg y \wedge \neg z) \vee (y \wedge u)$. We have two positive instances of y, and one negative instance. The number of instances is not important, but the fact that we have a positive and a negative instance is.

Thus, we will distinguish (as before) the positive arrow $x \to y$, the negative arrow $x \nrightarrow y$, and (new) the situation where we have a positive and a negative arrow from x to y, written as $x \to_{\pm} y$, see also Diagram 7.3.2 (page 139).

(3) Labelled path

Let σ be a path as usual in the graph $\mathcal{G}$.

σ_0, σ_1, etc. will be the nodes on the path.

(We are not very consistent here, sometimes σ_0, σ_1 etc. will also be different paths - context will tell.)

As the graph is free from cycles, σ_0, σ_n, σ_x, $\sigma_{<x}$, $\sigma_{\leq x}$, $\sigma_{>x}$, $\sigma_{\geq x}$ will all be defined if x is a node on the path. (The latter will indicate the path (or its elements) up to x without x, etc.)

We define the labelled path from the path σ inductively.

σ_0 will be the first element of the labelled path, too.

Let the labelled path be defined up to σ_n.

If $\sigma_n \to \sigma_{n+1}$, then $\sigma_0 \dots \sigma_n \to \sigma_{n+1}$ is the labelled path up to σ_{n+1}.

If $\sigma_n \not\to \sigma_{n+1}$, then $\sigma_0 \ldots \sigma_n \not\to \sigma_{n+1}$ is the labelled path up to σ_{n+1}.

If $\sigma_n \to_\pm \sigma_{n+1}$, then we split into two different labelled paths: $\sigma_0 \ldots \sigma_n \to \sigma_{n+1}$ is the first labelled path up to σ_{n+1}, and $\sigma_0 \ldots \sigma_n \not\to \sigma_{n+1}$ is the second labelled path up to σ_{n+1}.

Thus, $\to_\pm$ does not occur any more.

(4) Valued paths

We define the valued paths from a labelled path σ by induction from an initial value + or - for σ_0, by propagation.

Let $v(\sigma_0) := +$ or -, corresponding to the initial value.

Let $v(\sigma_0) \ldots v(\sigma_n)$ be defined.

(4.1) Case 1: $v(\sigma_n) = +$.
  If $\sigma_n \to \sigma_{n+1}$, then $v(\sigma_{n+1}) = +$,
  If $\sigma_n \not\to \sigma_{n+1}$, then $v(\sigma_{n+1}) = -$.

(4.2) Case 2: $v(\sigma_n) = -$.
  If $\sigma_n \to \sigma_{n+1}$, then $v(\sigma_{n+1}) = -$,
  If $\sigma_n \not\to \sigma_{n+1}$, then $v(\sigma_{n+1}) = +$.

Recall that $\sigma_n \to \sigma_{n+1}$ or $\sigma_n \not\to \sigma_{n+1}$, but never both in labelled paths.

(5) We denote σ^+ the valued path σ beginning with $v(\sigma_0) = +$, and σ^- the valued path σ beginning with $v(\sigma_0) = -$.

We call σ^+ the opposite of σ^-, and vice versa.

Fact 7.4.1

(1) If $v_1(\sigma_0)$ is the opposite of (contradicts) $v_2(\sigma_0)$, then so do $v_1(\sigma_n)$ to $v_2(\sigma_n)$ for all n.

(Proof by induction.)

(2) If $\sigma_{\geq x} = \tau_{\geq x}$ and $v(\sigma_x)$ contradicts $v'(\tau_x)$, then it also holds for all $y > x$.
(By (1).)

(3) If $v(\sigma_x)$ contradicts $v'(\tau_x)$, then $v^-(\sigma_x)$ contradicts $v'^-(\tau_x)$, too.

Example 7.4.1

Consider a graph with 5 nodes, x_0, x_1, x_2, x_3, x_4. x_4 will not have any successors. Let $d(x_0) = \neg x_1 \lor x_2 \lor x_4$, $d(x_1) = x_2$, $d(x_2) = x_3 \land \neg x_3$, $d(x_3) = x_4$. For brevity, we will sometimes also write $x_0 = \neg x_1 \lor x_2$ etc.

(1) The graph as in the Yablo structure is $x_0 \to x_1 \to x_2 \to x_3 \to x_4$, $x_0 \to x_2$, $x_0 \to x_4$.

(2) The graph with labelled arrows is $x_0 \nrightarrow x_1 \to x_2 \to_\pm x_3 \to x_4$, $x_0 \to x_2$, $x_0 \to x_4$.

(3) The labelled paths are

$\sigma^0 : x_0 \nrightarrow x_1 \to x_2 \to x_3 \to x_4$,

$\sigma^1 : x_0 \nrightarrow x_1 \to x_2 \nrightarrow x_3 \to x_4$,

$\sigma^2 : x_0 \to x_2 \to x_3 \to x_4$,

$\sigma^3 : x_0 \to x_2 \nrightarrow x_3 \to x_4$,

$\sigma^4 : x_0 \to x_4$.

(4) The valued paths are

(4.1) σ^+ :

$\sigma^{0+} : x_0^+ \nrightarrow x_1^- \to x_2^- \to x_3^- \to x_4^-$,

$\sigma^{1+} : x_0^+ \nrightarrow x_1^- \to x_2^- \nrightarrow x_3^+ \to x_4^+$,

$\sigma^{2+} : x_0^+ \to x_2^+ \to x_3^+ \to x_4^+$,

$\sigma^{3+} : x_0^+ \to x_2^+ \nrightarrow x_3^- \to x_4^-$,

$\sigma^{4+} : x_0^+ \to x_4^+$.

(4.2) σ^- :

$\sigma^{0-} : x_0^- \nrightarrow x_1^+ \to x_2^+ \to x_3^+ \to x_4^+$,

$\sigma^{1-} : x_0^- \nrightarrow x_1^+ \to x_2^+ \nrightarrow x_3^- \to x_4^-$,

$\sigma^{2-} : x_0^- \to x_2^- \to x_3^- \to x_4^-$,

$\sigma^{3-} : x_0^- \to x_2^- \nrightarrow x_3^+ \to x_4^+$,

$\sigma^{4-} : x_0^- \to x_4^-$.

7.4.2 Discussion of a Non-recursive $\bigvee \bigwedge$ Construction

Fact 7.4.2

(1) No triangles (this is a special case of (2))

Let ρ, σ, τ be labelled paths such that once they meet, they continue the same way. So, if ρ, σ meet at $x_{\rho,\sigma}$, then $\rho_{\geq x_{\rho,\sigma}} = \sigma_{\geq x_{\rho,\sigma}}$, etc.

Let $v_\rho(\rho)$, $v_\sigma(\sigma)$, $v_\tau(\tau)$ be valued versions of ρ, σ, τ.

Then it is impossible that $v_\rho(\rho_{x_{\rho,\sigma}})$ contradicts $v_\sigma(\sigma_{x_{\rho,\sigma}})$, $v_\sigma(\sigma_{x_{\sigma,\tau}})$ contradicts $v_\tau(\tau_{x_{\sigma,\tau}})$, $v_\tau(\tau_{x_{\tau,\rho}})$ contradicts $v_\rho(\rho_{x_{\tau,\rho}})$.

Case 1: $x_{\rho,\sigma} = x_{\sigma,\tau}$ ($= x_{\rho,\tau}$ by prerequisite). As we have only two values, this is impossible.

Case 2: Assume wlog that $x_{\rho,\sigma}$ is "above" $x_{\sigma,\tau}$. $v_\rho(\rho_{x_{\rho,\sigma}})$ contradicts $v_\sigma(\sigma_{x_{\rho,\sigma}})$ by prerequisite, but by (2) $v_\rho(\rho_{x_{\sigma,\tau}})$ contradicts $v_\sigma(\sigma_{x_{\sigma,\tau}})$, so $v_\tau(\tau_{x_{\sigma,\tau}})$ cannot contradict both $v_\rho(\rho_{x_{\sigma,\tau}})$ and $v_\sigma(\sigma_{x_{\sigma,\tau}})$, as we have only two values.

(2) **No loops of odd length**

We argue as in (3).

Let $\sigma^0, \ldots \sigma^{2n}$ be paths with valuations $v^0, \ldots v^{2n}$ such that $v^i(\sigma^i)$ contradicts $v^{i+1}(\sigma^{i+1})$ modulo $2n + 1$. This is impossible. Assume the contrary.

Let e.g. σ^0 meet σ^1 at x_{σ^0,σ^1}, and $v^0(\sigma^0_{x_{\sigma^0,\sigma^1}})$ contradict $v^1(\sigma^1_{x_{\sigma^0,\sigma^1}})$, and from x_{σ^0,σ^1} on, σ^0 and σ^1 are identical.

Argue for σ^1 and σ^2 the same way. $x_{\sigma^1\sigma^2}$ may be below x_{σ^0,σ^1}, above, or identical. When we consider $x_{\sigma^0\sigma^1\sigma^2} := min(x_{\sigma^0,\sigma^1}, x_{\sigma^1,\sigma^2})$, then $\sigma^0, \sigma^1, \sigma^2$ will be identical from $x_{\sigma^0\sigma^1\sigma^2}$ on, etc.

Finally, there is a point z where all σ^i are identical, and by (2), $v^i(\sigma^i_z)$ contradicts $v^{i+1}(\sigma^{i+1}_z)$, etc. Assume e.g. wlog. $v^0(\sigma^0_z) = +$, $v^1(\sigma^1_z) = -$, $v^2(\sigma^2_z) = +$, etc., and $v^{2n}(\sigma^{2n}_z) = +$, contradiction.

We may have arbitrarily many paths pairwise contradictory, as the following example shows. This is impossible within one cell.

Example 7.4.2

Let $\sigma_0 : x_0 \not\to x_1 \to x_2 \not\to x_3 \to x_4$, $\sigma_1 : x_0 \not\to x_1 \to x_2 \to x_4$, $\sigma_2 : x_0 \to x_2 \not\to x_3 \to x_4$, $\sigma_3 : x_0 \to x_2 \to x_4$,

then σ_0 contradicts σ_1 in the lower part, σ_2 and σ_3 in the upper part, σ_1 contradicts σ_2 and σ_3 in the upper part, σ_2 contradicts σ_3 in the lower part.

Obviously, this may be generalized to 2^ω paths.

The following example is for illustration, the general solution is in Proposition 7.4.3 (page 158).

See also Example 7.3.2 (page 133) for illustration.

Example 7.4.3

Suppose we work with DNF, and have solved the distributions, so $d(x) = \mathcal{A} \vee \mathcal{B} \vee \mathcal{C} \vee \mathcal{D}$, where $\mathcal{A} = \bigwedge\{a_i : i \in I_a\}$, $\mathcal{B} = \bigwedge\{b_i : i \in I_b\}$, $\mathcal{C} = \bigwedge\{c_i : i \in I_c\}$, $\mathcal{D} = \bigwedge\{d_i : i \in I_d\}$. As $d(x)$ is a disjunction, and, by prerequisite, $d(x)$ has to be contradictory, $\mathcal{A}$ etc. have to be contradictory for $x+$, thus there have to $A, a \in \{a_i : i \in I_a\}$ which are contradictory (recall: contradictions are always between two elements, see Fact 7.3.3 (page 133)). Likewise, there must be such B, b, C, c, D, d.

As shown in Fact 7.4.1 (page 154), A, a are contradictory, iff their opposites A^-, a^- are contradictory.

To make x- contradictory, every choice (of opposites) in $\mathcal{A}, \mathcal{B}, \mathcal{C}, \mathcal{D}$ has to contradictory (see Example 7.3.2 (page 133)), in particular, every choice in $\{A^-, a^-\}$, $(B^-, b^-\}$, $\{C^-, c^-\}$, $\{D^-, d^-\}$ has to be contradictory.

Thus,

$\{A^-, a^-\}$, $(B^-, b^-\}$, $\{C^-, c^-\}$, $\{D^-, d^-\}$ each are contradictory by prerequisite (for $x+$)

and all choice sets for x-

$\{A^-, B^-, C^-, D^-\}$, $\{A^-, B^-, C^-, d^-\}$, ..., $\{a^-, b^-, c^-, D^-\}$, $\{a^-, b^-, c^-, d^-\}$

have to be made contradictory.

We show that this leads to loops of contradictions of odd length, contradicting Fact 7.4.2 (page 155), (2). For notational simplicity, we denote A^- by A (or, using again Fact 7.4.1 (page 154)), so we have to chose a contradiction in all $\{A, B, C, D\}$, $\{A, B, C, d\}$, ..., $\{a, b, c, D\}$, $\{a, b, c, d\}$

See Diagram 7.4.1 (page 160).

The lines there connect contradictory couples.

The vertical lines are the contradictions originating from $x+$.

For reasons of symmetry, we may assume that $\{A, B\}$ are contradictory - recall again Fact 7.3.3 (page 133).

We may continue by adding the contradictions $\{A, C\}$, or $\{B, C\}$ as in the diagram, or $\{a, C\}$, or $\{C, D\}$. The other cases are symmetrical.

(1) Fix now $\{B, C\}$. The choices in $\{A, b, C, D\}$ and $\{A, b, C, d\}$ together are impossible, as we will show by examining the cases.

   (Note that in the set $\{A, b, C\}$ all nodes A, b, C have even distance from A. This is the basic reason for the fact that adding xD and $x'd$ leads to a cycle of uneven length: $A \ldots x D d x' \ldots A$. This is elaborated in Proposition 7.4.3 (page 158).)

   Chosing $\{A, b\}$, $\{A, C\}$, $\{b, C\}$ are all impossible, as they lead to loops of length 3, e.g. $A - b - B - A$.

   (We simplify notation.)

   The possible choices for AbCD are thus AD, bD, CD, and for AbCd Ad, bd, Cd.

   (1.1) Assume the choice AD for AbCD. Then the choice of Ad is impossible by the loop ADdA, the choice of bd by the loop ADdbBA, the choice of Cd by ADdCBA.

   (1.2) Assume bD for AbCD. Then Ad is impossible by AdDbBA, bd by bdDb, Cd by BCdDbB (or ABCdDbBA)

   (1.3) Assume CD for AbCD. Then Ad is impossible by AdDCBA, bd by bdDCBb, Cd by CDdC.

   So $\{B, C\}$ is impossible.

(2) Consider the case of $\{C, D\}$ added as contradictory: We show that AbCd and AbcD together are impossible.

   Ab, Cd, cD are impossible.

For AbCd, we consider AC, Ad, bC, bd as candidates for contradiction.

For AbcD, we consider Ac, AD, bc, bD as candidates for contradiction.

(2.1) Assume AC: Ac is impossible, AD because of ACDA, bc because of ABbcCA, bD because of ACDbBA.

(2.2) Assume Ad: AD is impossible, Ac because of AdDCcA, bc because of AdDCcbBA, bD because of AdDbBA.

(2.3) Assume bC: bc is impossible, AC because of AcCbBA, AD because of ADCbBA, bc because of bCDb.

(2.4) Assume bd: bD is impossible, Ac because of AcCDdbBA, AD because of ADdbBA, bc because of bdDCcb.

So $\{C, D\}$ is impossible.

(3) The other cases are similar.

We first checked this and other examples with a small computer program.

We make this more general.

Proposition 7.4.3

We cannot make both $x+$ and x- contradictory.

The proof goes over several steps. We will construct a sequence like ABcDef ...., dscribing the valuation A, B, c, D, e, f .... which will be consistent, i.e. there is no contradiction between A and c, etc., graphically no line A-c, and any attempt to make it inconsistent will result in a cycle of odd length, contradicting Fact 7.4.2 (page 155).

Definition 7.4.2

(1) (A, a) and (B, b) are directly connected iff $\{A, B\}$, or $\{A, b\}$, or $\{a, B\}$, or $\{a, b\}$ are contradictory, i.e., in our graphical notation, iff there is a line from A to B, or .....

(Recall than $\{A, a\}$ are contradictory for $x + .$)

(2) (A_0, a_0) and (A_n, a_n) are connected iff there is a sequence of directly connected pairs $(A_0, a_0), (A_1, a_1), \ldots, (A_n, a_n)$.

(3) A set of pairs $\mathcal{A} = \{(A_0, a_0), \ldots, (A_n, a_n)\}$ is connected iff all (A_0, a_0) and (A_i, a_i) are connected. (Instead of (A_0, a_0) any other pair will do.)

(4) A maximal connected set is called a bloc.

We assume now that all blocs are consistent (no odd loops of contradictions), see Fact 7.4.2 (page 155).

Fact 7.4.4

(1) Let $\mathcal{B}$ be a bloc. Fix $(A_0, a_0) \in \mathcal{B}$ arbitrarily. Fix A_0 (or a_0) arbitrarily. If $\mathcal{B} = \{(A_0, a_0)\}$, we are done with this bloc.

(2) Let $(A_0, a_0) \neq (X, x) \in \mathcal{B}$. Then there is a path of contradictions of even length from A_0 to X or x.

(3) Example:

Recall that $\mathcal{B}$ is connected. Let e.g. $A_0 - a_0 - a_1 - A_1 - A_2 - x$. This has length 5, but adding $x - X$ results in a path of even length 6.

We do this for all $(X, x) \in \mathcal{B}$ different from (A_0, a_0).

(4) This results in a set e.g. of $\mathcal{C} = \{A_0, X, x', x'', X''', \ldots\}$, where all X, x' etc. have a path of contradictions of even length from A_0 - but also from each other, we just have to go back via A_0.

(5) We call this set $\mathcal{C}$ an even choice set for $\mathcal{B}$, and write it $\{\beta_0, \beta_1, \ldots etc.\}$. Seen as a (partial) valuation, it is consistent, i.e. there are no direct lines linking e.g. $\beta_0 - \beta_1$, so $x+$ and $x-$ cannot both be contradictory.

(6) Note that adding any new contradiction $\beta_i - \beta_j$ results in an odd loop of contradictions, which cannot be, see Fact 7.4.2 (page 155).

(7) Suppose we have a set $\{(A_i, a_i) : i \in I\}$, composed of blocs $\mathcal{B}_j : j \in J$. Take now a valuation as above for each bloc $\mathcal{B}_j$. Then

(7.1) The union of all even choice sets is free from contradictions.

(7.2) We cannot add any contradiction within any one bloc - see above, (6).

(7.3) Thus, if we have just one bloc, this valuation cannot be made contradictory in a consistent way.

(7.4) We can make the union of choice sets contradictory in a consistent way by adding an inter-bloc contradiction.

(8) Thus, we have created a full valuation which is consistent, and any attempt to make it inconsistent will result in a cycle of odd length.

□ Proposition 7.4.3 (page 158)

Example 4-case, Case (1)

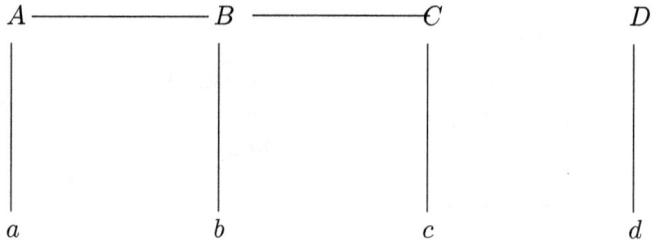

```
A ————————— B ————————— C          D
|           |           |          |
|           |           |          |
|           |           |          |
a           b           c          d
```

Example 4-case, Case (2)

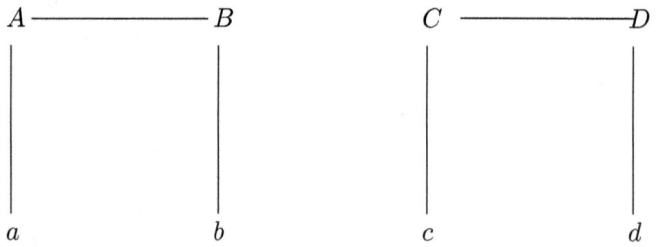

Diagram 7.4.1

```
A ————————— B          C ————————— D
|           |          |           |
|           |          |           |
|           |          |           |
a           b          c           d
```

7.5   A Generalization of Yablo's Construction to $\bigvee \bigwedge$

We discuss here a general strategy and a number of examples (in Example 7.5.1 (page 164)).

They have in common that the formulas are of the type $\bigvee \bigwedge$, i.e. in disjunctive normal form. The limiting cases are pure conjunctions (as in Yablo's original approach) or pure disjunctions.

The examples are straightforward generalizations of Yablo's construction, as we have here several columns, in Yablo's construction just one column, and our choice functions g (in all columns) correspond to the choice of one element in Yablo's construction.

Consider the first example below.

More precisely, as Yablo works with $\bigwedge x_i$, there is one uniform set of x_i. We work with $\bigvee \bigwedge x_{i,j}$, so we have to distinguish the elements in the $\bigwedge$ from the sets in $\bigvee$. We define for this purpose columns, whose elements are the elements in the $\bigwedge$, and the set of columns are the sets in $\bigvee$. Negation is now slightly more complicated, not just OR of negated elements, but OR of choice functions of negated elements in the columns. We also need some enumeration of $\omega \times \omega$, or of a suitable set.

It is perhaps easiest to see the following example geometrically. We have a vertical column C_{i_0} where a certain property holds, and a horizontal line (the choice function g), where the opposite holds, and column and horizontal line meet.

More precisely, the property will not necessarily hold in all of C_{i_0}, but only from a certain height onward. As the choice functions g will chose even higher in the columns, the clash is assured.

For more details of the general strategy, see Definition 7.5.1 (page 161), Case 5 (page 162). and Example 7.5.1 (page 164), in particular Case 6 (page 165).

We first introduce some notation and definitions

Definition 7.5.1

(1) The examples will differ in the size of the $\bigwedge$ and $\bigvee$ - where the case of both finite is trivial, no contradictions possible - and the relation in the graph, i.e. which formulas are "visible" from a given formula. (In Yablo's construction, the x_j, $j > i$, are visible from x_i.) We will write the relation by $\not\to$, $x \not\to y$, but for simplicity sometimes $x < y$, too, reminding us that variables inside the formulas will be negated. We consider linear and ranked orders, leaving general partial orders aside. All relations will be transitive and acyclic.

For the intuition (and beyond) we order the formulas in sets of columns. Inside a column, the formulas are connected by $\bigwedge$, the columns themselves are connected by $\bigvee$. This is the logical ordering, it is different from above order relation in the graph.

(2) The basic structure.

(2.1) So, we have columns C_i, and inside the columns variables $x_{i,j}$. As we might have finitely or countably infinitely many columns, of finite or countably infinite size, we write the set of columns $C := \{C_i : i < \alpha\}$, where $\alpha < \omega + 1$, and $C_i := \{x_{i,j} : j < \beta_i\}$, where $\beta_i < \omega + 1$ again. By abuse of language, C will also denote the whole construction, $C := \{x_{i,j} : i, j < \omega\}$.

(2.2) Given C and $x_{i,j}$, $C \upharpoonright x_{i,j} := \{x_{i',j'} : x_{i,j} \not\rightarrow x_{i',j'}\}$, the part of C visible from $x_{i,j}$.

(2.3) Likewise, $C_{i'} \upharpoonright x_{i,j} := \{x_{i',j'} \in C_{i'} : x_{i,j} \not\rightarrow x_{i',j'}\}$.

(3) Let $I(x_{i,j}) := \{i' < \alpha : C_{i'} \upharpoonright x_{i,j} \neq \emptyset\}$. (In some i', there might be no $x_{i',j'}$ s.t. $x_{i,j} \not\rightarrow x_{i',j'}$.)

Given $i' \in I(x_{i,j})$, let $J(i', x_{i,j}) := \{j' < \beta_{i'} : x_{i,j} \not\rightarrow x_{i',j'}\}$. By $i' \in I(x_{i,j})$, $J(i', x_{i,j}) \neq \emptyset$.

(4) Back to logic. Let $d(x_{i,j}) := \bigvee\{\bigwedge\{\neg x_{i',j'} : j' \in J(i', x_{i,j})\} : i' \in I(x_{i,j})\}$, ($\wedge$ inside columns, $\vee$ between columns). We will sometimes abbreviate $d(x_{i,j})$ by $x_{i,j}$. (Note that the $x_{i',j'}$ are exactly the elements visible from $x_{i,j}$.)

(4.1) If $x_{i,j}$ is true (written $x_{i,j}+$), then all elements in one of the $C_{i'}$, $i' \in I(x_{i,j})$, and visible from $x_{i,j}$, must all be false - but we do not know in which $C_{i'}$. We denote this $C_{i'}$ by $C_{i(x_{i,j})}$, and define $C[x_{i,j}] := C_{i(x_{i,j})} \upharpoonright x_{i,j}$.

(4.2) Conversely, suppose $x_{i,j}$ is false, $x_{i,j}-$. Again, we consider only elements in $C \upharpoonright x_{i,j}$, i.e. visible from $x_{i,j}$. By distributivity, $x_{i,j}- = \bigvee\{ran(g) : g \in \Pi\{C_{i'} \upharpoonright x_{i,j} : i' \in I(x_{i,j})\}\}$.

(g is a choice function chosing in all columns $C_{i'}$ the "sufficiently big" elements $x_{k,m}$, i.e. above $x_{i,j}$, $ran(g)$ its range or image.)

Note that the elements of $ran(g)$ are now positive!

The $\bigvee$ in above formula choses some such function g, but we do not know which. Let $g[x_{i,j}]$ denote the chosen one.

(4.3) Note that both $C[x_{i,j}]$ and $g[x_{i,j}]$ are undefined if there are no $x_{i',j'} > x_{i,j}$.

(5) The conflicts will be between the "vertical" (negative) columns, and "horizontal" (positive) lines of the g. It is a very graphical construction. If we start with a positive point, the negative columns correspond to the $\forall$ in Yablo's construction, the g to the $\exists$, if we start with a negative one, it is the other way round.

More precisely:

(5.1) Let $x_{i,j}+$, consider $C[x_{i,j}] \subseteq C_{i(x_{i,j})}$ - recall all elements of $C[x_{i,j}]$ are negative.

If there is $x_{i',j'} \in C[x_{i,j}]$ which is not maximal in $C[x_{i,j}]$, then $g[x_{i',j'}]$ intersects $C[x_{i,j}]$, a contradiction, as all elements of $ran(g[x_{i',j'}])$ are positive.

Of course, such non-maximal $x_{i',j'}$ need not exist. In that case, we have to try again with the negative element $x_{i',j'}$ and Case (5.2).

(5.2) Let $x_{i,j}-$, consider $g[x_{i,j}]$ - recall, all elements of $ran(g[x_{i,j}])$ are positive.

If there is $x_{i'j'} \in ran(g[x_{i,j}])$ such that $C[x_{i',j'}] \cap ran(g[x_{i,j}]) \neq \emptyset$, we have a contradiction. (This case is impossible in Yablo's original construction.)

Otherwise, we have to try to work with the elements in $ran(g[x_{i,j}])$ and Case (5.1) above, etc.

Note that we can chose a suitable $x_{i',j'} \in ran(g[x_{i,j}])$, in particular one which is not maximal (if such exist), but we have no control over the choice of $C[x_{i',j'}]$, so we might have to exhaust all finite columns, until the choice is only from a set of infinite columns.

See Diagram 7.5.1 (page 166) and Example 7.5.1 (page 164), Case 6 (page 165).

Example 7.5.1

(1) Case 1

Consider the structure $C := \{x_{i,j} : i < \alpha, j < \omega\}$, with columns $C_i := \{x_{i,j} : j < \omega\}$.

Take a standard enumeration f of C, e.g. $f(0) := x_{0,0}$, then enumerate the $x_{i,j}$ s.t. $max\{i,j\} = 1$, then $max\{i,j\} = 2$, etc. As f is bijective, $f^{-1}(x_{i,j})$ is defined.

(More precisely, let $m := max\{i,j\}$, we go first horizontally from left to right over the columns up to column $m - 1$, then in column m upwards, i.e. $x_{0,m}, \ldots, x_{m-1,m}, x_{m,0}, \ldots, x_{m,m}$.)

Define the relation $x_{i,j} \not\rightarrow x_{i',j'}$ iff $f^{-1}(x_{i,j}) < f^{-1}(x_{i',j'})$. Obviously, $\not\rightarrow$ is transitive and free from cycles. $C \restriction k := \{x_{i,j} \in C : f^{-1}(x_{i,j}) > k\}$ etc. are defined.

We now show that the structure has no truth values.

Suppose $x_{i,j} +$.

Consider $C[x_{i,j}]$, let $i' := i(x_{i,j})$ and chose $x_{i',j'} \in C[x_{i,j}]$. $x_{i',j'}$ is false, $ran(g[x_{i',j'}])$ intersects $C_{i'}$ above $x_{i',j'}$, so we have a contradiction.

In particular, $x_{0,0}+$ is impossible.

Suppose $x_{0,0}-$, then $ran(g[x_{0,0}]) \neq \emptyset$, chose $x_{i,j} \in ran(g[x_{0,0}])$, so $x_{i,j}+$, but we saw that this is impossible.

Note: for $\alpha = 1$, we have Yablo's construction.

(2) Case 2

We now show that the same construction with columns of height 2 does not work, it has an escape path.

Set $x_{i,0}+$, $x_{i,1}-$ for all i. $C_{i(x_{i,0})}$ might be C_i, so $C[x_{i,0}] = \{x_{i,1}\}$, which is possible. $ran(g[x_{i,1}])$ might be $\{x_{j,0} : j > i\}$, which is possible again.

(3) Case 3

We change the order in Case 2 (page 164), to a (horizontally) ranked order: $x_{i,j} < x_{i',j'}$ iff $i < i'$ (thus, between columns), we now have a contradiction:

Consider $x_{i,0}-$ and $g[x_{i,0}]$. $g[x_{i,0}](i)$ is undefined. Let $x_{i+1,j'} := g[x_{i,0}](i+1)$, thus $x_{i+1,j'} +$. Consider $C[x_{i+1,j'}]$. This must be some $C_{i'}$ for $i' > i + 1$, so $C[x_{i+1,j'}] \cap ran(g[x_{i,0}]) \neq \emptyset$, and we have a contradiction.

For $x_{0,0}+$, consider $C[x_{0,0}]$, this must be some C_i, $i > 0$, take $x_{i,0} \in C_i$, $x_{i,0}-$, and continue as above.

(4) Case 4

Modify Case 1 (page 164), to a ranked order, but this time horizontally: $x_{i,j} < x_{i',j'}$ iff $j < j'$.

Take $x_{i,j}+$, consider $C[x_{i,j}]$, this may be (part of) any $C_{i'}$ (beginning at $j + 1$). Take e.g. $x_{i',j+1}- \in C[x_{i,j}]$, consider $g[x_{i',j+1}]$, a choice function in $\{C_k \upharpoonright x_{i',j+1} : k, \omega\}$ this will intersect $C[x_{i,j}]$. In particular, $x_{0,0}+$ is impossible.

Suppose $x_{0,0}-$, take $x_{i,j}+ \in ran(g[x_{0,0}])$, and continue as above.

Note that the case with just one C_i is the original Yablo construction.

(5) Case 5

We modify Case 1 (page 164) again: Consider the ranked order by $x_{i,j} \not\rightarrow x_{i',j'}$ iff $max\{i, j\} < max\{i', j'\}$.

This is left to the reader as an exercise.

(6) Case 6

The general argument is as follows (and applies to general partial orders, too):

(6.1) We show that $x_{i,j}+$ leads to a contradiction. (In our terminology, $x_{i,j}$ is the head.)

(6.1.1) we find $C[x_{i,j}]$ (negative elements) above $x_{i,j}$ - but we have no control over the choice of $C[x_{i,j}]$,

(6.1.2) we chose $x_{i',j'} \in C[x_{i,j}]$ - if there is a minimal such, chose this one, it must not be a maximal element in $C[x_{i,j}]$ ($x_{i',j'}$ is the knee),

(6.1.3) we find $g[x_{i',j'}]$ (positive elements) above $x_{i',j'}$ - again we have no control over the choice of this g. But, as $C[x_{i,j}] \upharpoonright x_{i',j'}$ is not empty, $ran(g[x_{i',j'}]) \cap C[x_{i,j}] \neq \emptyset$, so we have a contradiction ($x_{i'',j''} \in ran(g[x_{i',j'}]) \cap C[x_{i,j}]$ is the foot).

(6.1.4) We apply the reasoning to $x_{0,0}$.

(6.2) We show that $x_{0,0}-$ leads to a contradiction.

(6.2.1) We have $g[x_{0,0}]$ (positive elements) - but no control over the choice of g. In particular, it may be arbitrarily high up.

(6.2.2) we chose $x_{i,j} \in ran(g[x_{0,0}])$ with enough room above it for the argument about $x_{i,j}+$ in (6.1).

(6.3) Case 3 (page 164) is similar, except that we work horizontally, not vertically.

Diagram for Definition 7.5.1 Case 5

Case 5, (4.1)

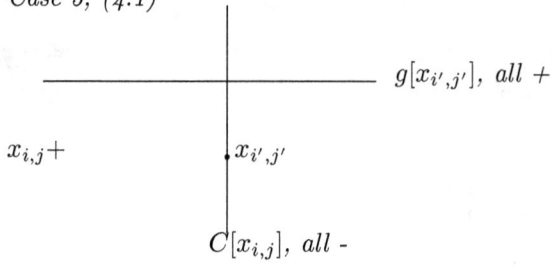

$g[x_{i',j'}]$, *all* $+$

$x_{i,j}+$

$x_{i',j'}$

$C[x_{i,j}]$, *all* -

Case 5, (4.2)

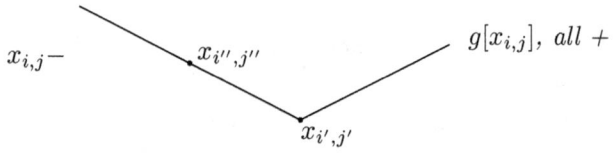

$x_{i,j}-$

$x_{i'',j''}$

$g[x_{i,j}]$, *all* $+$

$x_{i',j'}$

Diagram 7.5.1

Diagram for Example 7.5.1 Case 1

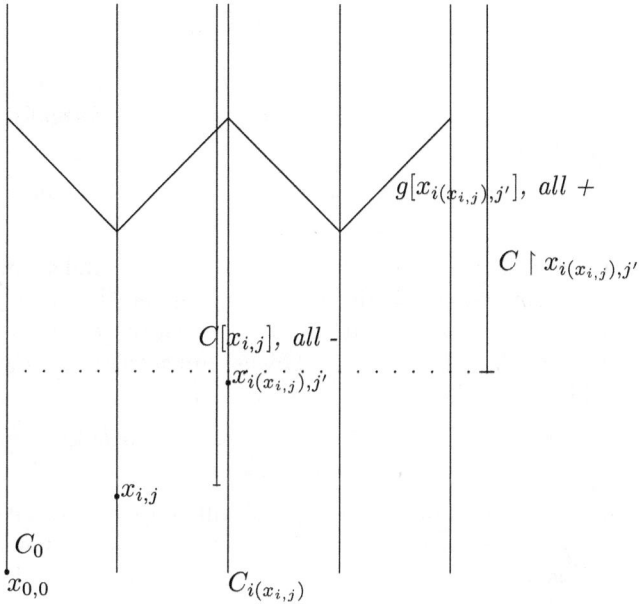

Diagram 7.5.2

7.6   Saw Blades

7.6.1   Introduction

Yablo works with contradictions in the form of $x_0 = \neg x_1 \wedge \neg x_2$, $x_1 = \neg x_2$, graphically $x_0 \nrightarrow x_1 \nrightarrow x_2$, $x_0 \nrightarrow x_2$. They are combined in a formally simple total order, which, however, blurs conceptual differences.

We discuss here different, conceptually very clear and simple, examples of a Yablo-Like construction.

In particular, we emphasize the difference between x_1 and x_2 in the Yablo contradictions. The contradiction is finished in x_2, but not in x_1, requiring the barring of escape routes in x_1. We repair the possible escape routes by constructing new contradictions for the SAME origin. This is equivalent to closing under transitivity in the individual "saw blades" - see below.

So we use the same "cells" as Yablo does for the contradictions, but analyse the way they are put together.

We use the full strength of the conceptual difference between x_1 and x_2 (in above notation) only in Section 7.6.4 (page 177), where we show that preventing x_2 from being TRUE is sufficient, whereas we need x_1 to be contradictory, see also Fact 7.6.3 (page 173), and Remark 7.6.4 (page 176). Thus, we obtain a minimal, i.e. necessary and sufficient, construction for combining Yablo cells in this way.

7.6.2   Saw Blades

First, we show the escape route problem.

Example 7.6.1

Consider Construction 7.6.1 (page 168) without closing under transitivity, i.e. the only arrows originating in $x_{\sigma,0}$ will be $x_{\sigma,0} \nrightarrow x_{\sigma,1}$ and $x_{\sigma,0} \nrightarrow y_{\sigma,0}$, etc.

Let $x_{\sigma,0} = TRUE$, then $x_{\sigma,1}$ is an $\vee$, and we pursue the path $x_{\sigma,0} \nrightarrow x_{\sigma,1} \nrightarrow x_{\sigma,2}$, this has no contradiction so far, and we continue with $x_{\sigma,2} = TRUE$, $x_{\sigma,3}$ is $\vee$ again, we continue and have $x_{\sigma,0} \nrightarrow x_{\sigma,1} \nrightarrow x_{\sigma,2} \nrightarrow x_{\sigma,3} \nrightarrow x_{\sigma,4}$, etc., never meeting a contradiction, so we have an escape path.

Construction 7.6.1

We construct a saw blade σ, SB_σ.

(1) "Saw Blades"

(1.1) Let $x_{\sigma,0} \nrightarrow x_{\sigma,1} \nrightarrow x_{\sigma,2} \nrightarrow x_{\sigma,3} \nrightarrow x_{\sigma,4}, \ldots$.
$x_{\sigma,0} \nrightarrow y_{\sigma,0}$, $x_{\sigma,1} \nrightarrow y_{\sigma,0}$, $x_{\sigma,1} \nrightarrow y_{\sigma,1}$, $x_{\sigma,2} \nrightarrow y_{\sigma,1}$, $x_{\sigma,2} \nrightarrow y_{\sigma,2}$, $x_{\sigma,3} \nrightarrow y_{\sigma,2}$, $x_{\sigma,3} \nrightarrow y_{\sigma,3}$, $x_{\sigma,4} \nrightarrow y_{\sigma,3}$, $\ldots$.

we call the construction a "saw blade", with "teeth" $y_{\sigma,0}$, $y_{\sigma,1}$, $y_{\sigma,2}$, ... and "back" $x_{\sigma,0}$, $x_{\sigma,1}$, $x_{\sigma,2}$, ...

We call $x_{\sigma,0}$ the start of the blade.

See Diagram 7.6.1 (page 171).

(1.2) Add (against escape), e.g. first $x_{\sigma,0} \not\to x_{\sigma,2}$, $x_{\sigma,0} \not\to y_{\sigma,1}$, then $x_{\sigma,1} \not\to x_{\sigma,3}$, $x_{\sigma,1} \not\to y_{\sigma,2}$, now we have to add $x_{\sigma,0} \not\to x_{\sigma,3}$, $x_{\sigma,0} \not\to y_{\sigma,2}$, etc, recursively. This is equivalent to closing the saw blade under transitivity with negative arrows $\not\to$. This is easily seen.

(1.3) We define the valuation by $d(x_{\sigma,i}) := \bigwedge \neg z_{\sigma,j}$, for all $x_{\sigma,j}$ such that $x_{\sigma,i} \not\to z_{\sigma,j}$, as in the original Yablo construction.

(2) Composition of saw blades

(2.1) Add for the teeth $y_{\sigma,0}$, $y_{\sigma,1}$, $y_{\sigma,2}$ .... their own saw blades, i.e. start at $y_{\sigma,0}$ a new saw blade $SB_{\sigma,0}$ with $y_{\sigma,0} = x_{\sigma,0,0}$, at $y_{\sigma,1}$ a new saw blade $SB_{\sigma,1}$ with $y_{\sigma,1} = x_{\sigma,1,0}$, etc.

(2.2) Do this recursively.

I.e., at every tooth of every saw blade start a new saw blade. See Diagram 7.6.2 (page 172).

Note:

It is NOT necessary to close the whole structure (the individual saw blades together) under transitivity.

Fact 7.6.1

All $z_{\sigma,i}$ in all saw blades so constructed are contradictory, i.e. assigning them a truth value leads to a contradiction.

Proof

Fix some saw blade SB_{σ} in the construction.

(1) Take any $z_{\sigma,i}$ with $z_{\sigma,i}+$, i.e. $z_{\sigma,i} = TRUE$. We show that this is contradictory.

(1.1) Case 1: $z_{\sigma,i}$ is one of the $x_{\sigma,i}$ i.e. it is in the back of the blade.

Take any $x_{\sigma,i'}$ in the back such that there is an arrow $x_{\sigma,i} \not\to x_{\sigma,i'}$ ($i' := i + 1$ suffices). Then $x_{\sigma,i'} = FALSE$, and we have an $\vee$ at $x_{\sigma,i'}$. Take any $z_{\sigma,j}$ such that $x_{\sigma,i'} \not\to z_{\sigma,j}$, by transitivity, $x_{\sigma,i} \not\to z_{\sigma,j}$, so $z_{\sigma,j} = FALSE$, but as $x_{\sigma,i'} = FALSE$, $z_{\sigma,j} = TRUE$, contradiction.

(1.2) Case 2: $z_{\sigma,i}$ is one of the $y_{\sigma,i}$, i.e. a tooth of the blade.

Then $y_{\sigma,i}$ is the start of the new blade starting at $y_{\sigma,i}$, and we argue as above in Case 1.

(2) Take any $z_{\sigma,i}$ with $z_{\sigma,i}-$, i.e. $z_{\sigma,i} = FALSE$, and we have an $\vee$ at $z_{\sigma,i}$, and one of the successors of $z_{\sigma,i}$, say $z_{\sigma,j}$, has to be TRUE. We just saw that this is impossible.

(For the intuition: If $z_{\sigma,i}$ is in the back of the blade, all of its successors are in the same blade. If $z_{\sigma,i}$ is one of the teeth of the blade, all of its successors are in the new blade, starting at $z_{\sigma,i}$. In both cases, $z_{\sigma,j} = TRUE$ leads to a contradiction, as we saw above.)

Remark 7.6.2

Note that all $y_{\sigma,i}$ are contradictory, too, not only the $x_{\sigma,i}$. We will see in Section 7.6.4 (page 177) that we can achieve this by simpler means, as we need to consider here the case $x_{\sigma,i}\vee$ only, the contradiction for the case $x_{\sigma,i}\wedge$ is already treated.

Thus, we seemingly did not fully use here the conceptual clarity of difference between x_1 and x_2 alluded to in the beginning of Section 7.6.2 (page 168). See, however, the discussion in Section 7.6.3.2 (page 173).

Diagram Single Saw Blade
Start of the saw blade σ beginning at $x_{\sigma,0}$,
before closing under transitivity

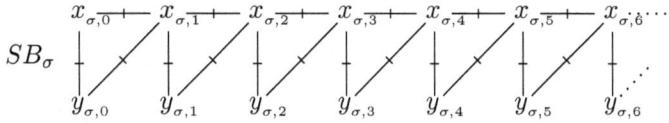

Diagram 7.6.1

Read $x_{\sigma,0} \not\to x_{\sigma,1} \not\to y_{\sigma,0}$, $x_{\sigma,0} \not\to y_{\sigma,0}$, etc, more precisely $x_{\sigma,0} = \neg x_{\sigma,1} \wedge \neg y_{\sigma,0}$, $x_{\sigma,1} = \neg y_{\sigma,0} \wedge \neg x_{\sigma,2} \wedge \neg y_{\sigma,1}$, etc.

Diagram Composition of Saw Blades
Composition of saw blades (without additional arrows)
The fat dots indicate identity, e.g. $y_{0,0} = x_{0,0,0}$

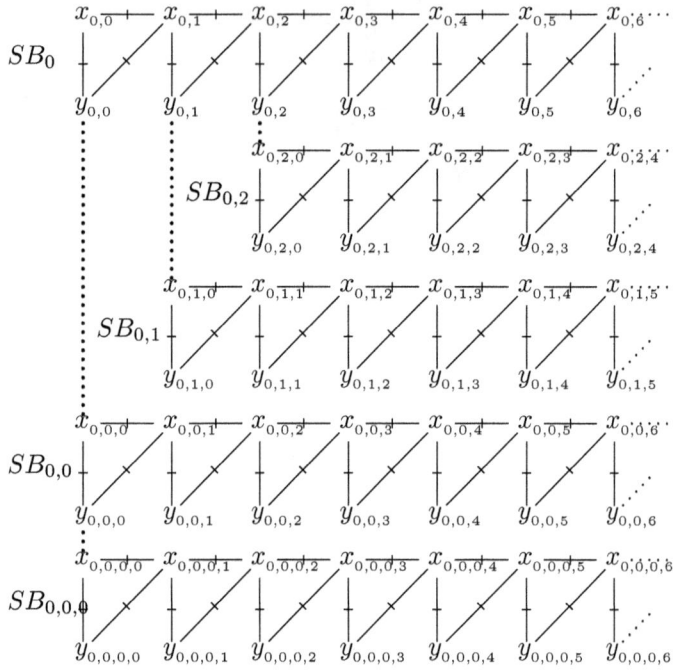

$$x_{0,0} \quad x_{0,1} \quad x_{0,2} \quad x_{0,3} \quad x_{0,4} \quad x_{0,5} \quad x_{0,6} \cdots$$

SB_0

$$y_{0,0} \quad y_{0,1} \quad y_{0,2} \quad y_{0,3} \quad y_{0,4} \quad y_{0,5} \quad y_{0,6}$$

$$x_{0,2,0} \quad x_{0,2,1} \quad x_{0,2,2} \quad x_{0,2,3} \quad x_{0,2,4} \cdots$$

$SB_{0,2}$

$$y_{0,2,0} \quad y_{0,2,1} \quad y_{0,2,2} \quad y_{0,2,3} \quad y_{0,2,4}$$

$$x_{0,1,0} \quad x_{0,1,1} \quad x_{0,1,2} \quad x_{0,1,3} \quad x_{0,1,4} \quad x_{0,1,5} \cdots$$

$SB_{0,1}$

$$y_{0,1,0} \quad y_{0,1,1} \quad y_{0,1,2} \quad y_{0,1,3} \quad y_{0,1,4} \quad y_{0,1,5}$$

$$x_{0,0,0} \quad x_{0,0,1} \quad x_{0,0,2} \quad x_{0,0,3} \quad x_{0,0,4} \quad x_{0,0,5} \quad x_{0,0,6} \cdots$$

$SB_{0,0}$

$$y_{0,0,0} \quad y_{0,0,1} \quad y_{0,0,2} \quad y_{0,0,3} \quad y_{0,0,4} \quad y_{0,0,5} \quad y_{0,0,6}$$

$$x_{0,0,0,0} \quad x_{0,0,0,1} \quad x_{0,0,0,2} \quad x_{0,0,0,3} \quad x_{0,0,0,4} \quad x_{0,0,0,5} \quad x_{0,0,0,6}$$

$SB_{0,0,0}$

$$y_{0,0,0,0} \quad y_{0,0,0,1} \quad y_{0,0,0,2} \quad y_{0,0,0,3} \quad y_{0,0,0,4} \quad y_{0,0,0,5} \quad y_{0,0,0,6}$$

Diagram 7.6.2

7.6.3 Discussion of Saw Blades

7.6.3.1 Simplifications

In general, and this does not only concern Saw Blade like constructions:

(1) If, say, x can be given the value TRUE (it has no successors, $x = y \vee \neg y$, etc.), then we can simplify a variable z where x occurs. If $z = x \wedge x'$, then we may set $z = x'$, if $z = z \vee x'$, we may set $z = TRUE$, etc.

(2) If the structure below x is a tree (no branches meet again), then we have no contradictions.

(3) We may contract finitely many branching points to one branching point, with equivalent structures (trivial), but not necessarily infinitely many branching points, see Example 7.3.7 (page 143).

Simplifications that will not work We try to simplify here the Saw Blade construction. Throughout, we consider formulas of pure conjunctions.

We start with a Yablo Cell, but try to continue otherwise.

So we have $x_0 \nrightarrow x_1 \nrightarrow x_2$, $x_1 \nrightarrow x_2$. So x_0+ is impossible. We now try to treat $x_0 -$. We see in Section 7.6.4 (page 177) that appending $x_2 \Rightarrow_{\pm} y_2$ may take care of the necessary contradiction at x_2, see Diagram 7.6.3 (page 179). When we try to do the same at x_1, i.e. some $x_1 \Rightarrow_{\pm} x_3$, we solve again the necessary contradiction at x_1, but run into a problem with x_0+, as x_1 is an $\vee$. So x_3 has to be contradictory. If we continue $x_3 \Rightarrow_{\pm} x_4 \Rightarrow_{\pm} x_5$ etc., this will not work, as we may set all such x_i-, and have a model. In abstract terms, we only procrastinate the same problem without solving anything. Of course, we could append after some time new Yablo Cells, as in the saw blade construction, but this is cheating, as the "true" construction begins only later.

Suppose we add not only $x_1 \Rightarrow_{\pm} x_3$, but also $x_0 \Rightarrow_{\pm} x_3$, then we solve x_0+, but x_0- is not solved.

Working with cells of the type (2.1) in Example 7.3.4 (page 136) will lead to similar problems.

Consequently, any attempt to use a "pipeline", avoiding infinite branching, is doomed:

Instead of $x_0 \nrightarrow x_1$, $x_0 \nrightarrow x_2$, ... etc. we construct a "pipeline" of x'_i, with $x_0 \nrightarrow x'_1$, $x_1 \nrightarrow x'_2$, etc, $x'_1 \rightarrow x'_2 \rightarrow x'_3$...., and $x'_1 \rightarrow x_1$, $x'_2 \rightarrow x_2$, etc. or similarly, to have infinitely many contradictions for paths from x_0.

As this is a set of classical formulas, this cannot achieve inconsistency, see Fact 7.1.2 (page 121).

7.6.3.2 Infinite Branching and Recursive Contradictions are Necessary

Fact 7.6.3

We need infinite branching in the saw blade construction at all x_i.

(We always use basic contradictions of the type $x_i \not\to x_{i+1} \not\to y_i$, $x_i \not\to y_i$ - which we abbreviate ∇.)

Proof

(1) The argument

(1.1) This is needed for x_0 only.

If x_0- is impossible, then all arrows $x_0 \not\to x_i$ have to lead to ∇ attached at x_i, so x_i+ must be impossible (recall x_0- means $\vee$ at x_0).

(Suppose we try to stop at the first ∇, then $x_1 \to y_0$ leads in hindsight to a contradiction, as it is again the x_0 of a new saw blade, but as it has value -, we run into a circularity, we still have to show that this is contradictory. For x_0+, this is different, we have to consider just one saw blade to see that this is contradictory.)

Thus, if $x_0 \not\to x_1$, $x_0 \not\to x_i$ exist, we attach at x_1 and x_i an ∇, e.g. $x_i \not\to y_i$, $x_i \not\to x_{i+1} \not\to y_i$, etc.

This is important (and possible at this level of analysis) only for x_0.

(1.2) The following holds for all i.

x_i+ impossible $\Rightarrow$ (by $x_i \not\to x_{i+1}$) all $x_{i+1} \not\to x_j$ and $x_{i+1} \not\to y_j$ must have a contradiction $x_i \not\to x_j$ viz. $x_i \not\to y_j$, so we have new arrows originating at x_i.

We need a branching at x_{i+1}, see Example 7.6.2 (page 180), Case (2.1). This holds for all i.

(2) So we have the following construction:

(2.1) start with ∇ $x_0 \not\to y_0$, $x_0 \not\to x_1 \not\to y_0$.

- So x_1 is the knee in the cell $x_0 \not\to y_0$, $x_0 \not\to x_1 \not\to y_0$.

(2.2) x_1

(2.2.1) as x_0- should be impossible, append a new ∇ to x_1, so x_1+ is impossible:

$x_0 \not\to x_1 \not\to x_2$, $x_0 \not\to y_0$, $x_1 \not\to y_0$, $x_1 \not\to y_1$, $x_2 \not\to y_1$.

From now on, we will not mention all y_i, only all x_i.

(2.2.2) as x_0+ should be impossible, add $x_0 \not\to x_2$ because of $x_1 \not\to x_2$, and $x_0 \not\to y_1$ because of $x_1 \not\to y_1$,

so we have $x_0 \not\to x_1 \not\to x_2$, $x_0 \not\to x_2$, and the new arrow $x_0 \not\to y_1$

- So x_1 is the knee for $x_0 \not\to x_1 \not\to x_2$, $x_0 \not\to x_2$, too, and x_2 its foot.

(2.3) x_2

(2.3.1) as x_0- should be impossible, append ∇ to x_2, because of the new arrow $x_0 \not\to x_2$, and x_2+ is impossible,
so we have $x_0 \not\to x_1 \not\to x_2 \not\to x_3$, $x_0 \not\to x_2$ and the new arrows $\not\to y_j$

(2.3.2) as x_1+ should be impossible, add $x_1 \not\to x_3$ because of $x_2 \not\to x_3$, and $x_1 \not\to y_2$ because of $x_2 \not\to y_2$,
so we have $x_0 \not\to x_1 \not\to x_2 \not\to x_3$, $x_0 \not\to x_2$, $x_1 \not\to x_3$ and the new arrows $\not\to y_j$

- So, for $x_1 \not\to x_2 \not\to x_3$, $x_1 \not\to x_3$ x_1 is the head, x_2 the knee, x_3 the foot.

- But, also, by $x_0 \not\to x_2 \not\to x_3$, $x_0 \not\to x_3$, here x_0 is the head and x_2 the knee, whereas in (2.2.1), x_0 was the head, and x_2 the foot.

(2.3.3) as x_0+ should be impossible, add $x_0 \not\to x_3$ because of $x_1 \not\to x_3$, and $x_0 \not\to y_2$ because of $x_1 \not\to y_2$,
so we have $x_0 \not\to x_1 \not\to x_2 \not\to x_3$, $x_0 \not\to x_2$, $x_1 \not\to x_3$, $x_0 \not\to x_3$, and the new arrows $\not\to y_j$

- So, for $x_0 \not\to x_1 \not\to x_3$, $x_0 \not\to x_3$, x_0 is the head, x_1 the knee, x_3 the foot. Etc.

(2.4) x_3

(2.4.1) as x_0- should be impossible, append ∇ to x_3, because of the new arrow $x_0 \not\to x_3$, and x_3+ is impossible,
so we have $x_0 \not\to x_1 \not\to x_2 \not\to x_3 \not\to x_4$, $x_0 \not\to x_2$, $x_1 \not\to x_3$, $x_0 \not\to x_3$ and the new arrows $\not\to y_j$

(2.4.2) as x_2+ should be impossible, add $x_2 \not\to x_4$ because of $x_3 \not\to x_4$, and $x_2 \not\to y_3$ because of $x_3 \not\to y_3$,
so we have $x_0 \not\to x_1 \not\to x_2 \not\to x_3 \not\to x_4$, $x_0 \not\to x_2$, $x_1 \not\to x_3$, $x_0 \not\to x_3$, $x_2 \not\to x_4$ and the new arrows $\not\to y_j$

(2.4.3) as x_1+ should be impossible, add $x_1 \not\to x_4$ because of $x_2 \not\to x_4$, and $x_1 \not\to y_3$ because of $x_2 \not\to y_3$,
so we have $x_0 \not\to x_1 \not\to x_2 \not\to x_3 \not\to x_4$, $x_0 \not\to x_2$, $x_1 \not\to x_3$, $x_0 \not\to x_3$, $x_2 \not\to x_4$, $x_1 \not\to x_4$ and the new arrows $\not\to y_j$

(2.4.4) as x_0+ should be impossible, add $x_0 \not\to x_4$ because of $x_1 \not\to x_4$, and $x_0 \not\to y_3$ because of $x_1 \not\to y_3$,
so we have $x_0 \not\to x_1 \not\to x_2 \not\to x_3 \not\to x_4$, $x_0 \not\to x_2$, $x_1 \not\to x_3$, $x_0 \not\to x_3$, $x_2 \not\to x_4$, $x_1 \not\to x_4$, $x_0 \not\to x_4$ and the new arrows $\not\to y_j$

(2.5) so we have a new arrow $x_0 \not\to x_4$, and apply again x_0-, etc.

(3) We see that the roles in the back of the saw blade change. x_1 begins as a knee, x_2 as a foot, later x_1 is a head, x_2 a knee, x_3 is a foot, etc. Whereas it is simple to treat feet (see Diagram 7.6.3 (page 179)), treating knees is more complicated, see also Section 7.6.3.1 (page 173). As the same nodes

change roles, we cannot have a "pure" construction according to our analysis (separate treatment for knees and feet). It seems difficult to separate the roles in a more complicated construction, e.g. by working with mixed $\vee\wedge$-formulas, see in particular cases (2.2.1) and (2.3.2) where x_0 is the head in both cases, x_2 the foot in one, the knee in another.

(4) Abstractly, we add complications in the steps (2.i), and repair them in the steps (2.i.j), so, in the limit, all damage done will be repaired. This is different from procrastination, where the same problem is just pushed to the future.

Thus, we have infinite branching for this construction.

$\square$.

The following remark shows that the construction has contradictory truth values recursively often.

Remark 7.6.4

We have a descending sequence of contradictory x_i - i.e. without attributable truth value - recursively often.

There are at least two arguments to show this:

- From the outside: otherwise, we could fill in truth values from the bottom.

- From the inside: we have to construct ever deeper ∇'s for x_0 to prevent escape paths, see steps (2.i.1) in Fact 7.6.3 (page 173).

7.6.4 Simplifications of the Saw Blade Construction

We show here that it is not necessary to make the $y_{\sigma,i}$ contradictory in a recursive construction, as in Construction 7.6.1 (page 168). It suffices to prevent them to be true.

We discuss three, much simplified, Saw Blade constructions.

Thus, we fully use here the conceptual difference of x_1 and x_2, as alluded to at the beginning of Section 7.6.2 (page 168).

Note, however, that the back of each saw blade "hides" a Yablo construction. The separate treatment of the teeth illustrates the conceptual difference, but it cannot escape blurring it again in the back of the blade.

See also Example 7.3.4 (page 136) for examples of simple contradictions.

Construction 7.6.2

(1) Take ONE saw blade σ, and attach (after closing under transitivity) at all $y_{\sigma,i}$ a SINGLE Yablo Cell $y_{\sigma,i} \nrightarrow u_{\sigma,i} \nrightarrow v_{\sigma,i}, y_{\sigma,i} \nrightarrow v_{\sigma,i}$. We call this the decoration, it is not involved in closure under transitivity.

   (1.1) Any node z in the saw blade (back or tooth) cannot have $z+$, this leads to a contradiction:

      If $z = x_i$ (in the back):

      Let x_i+ : Take $x_i \nrightarrow x_{i+1}$ (any x_j, $i < j$ would do), if $x_{i+1} \nrightarrow r$, then by transitivity, $x_i \nrightarrow r$, so we have a contradiction.

      If $z = y_i$ (a tooth):

      y_i+ is contradictory by the "decoration" appended to y_i.

   (1.2) Any x_i- (x_i in the back, as a matter of fact, x_0- would suffice) is impossible:

      Consider any $x_i \nrightarrow r$, then $r+$ is impossible, as we just saw.

      Note: there are no arrows from the back of the blade to the decoration.

(2) We can simplify even further. The only thing we need about the y_i is that they cannot be $+$. Instead of decorating them with a Yablo Cell, any contradiction will do, the simplest one is $y_i = y'_i \wedge \neg y'_i$. Even just one y' s.t. $y_i = y' \wedge \neg y'$ for all y_i would do. (Or a constant FALSE.)

See Diagram 7.6.3 (page 179).

Formally, we set

$$x_0 := \bigwedge\{\neg x_i : i > 0\} \wedge \bigwedge\{\neg y_i : i \geq 0\},$$

for $j > 0$:

$$x_j := \bigwedge\{\neg x_i : i > j\} \wedge \bigwedge\{\neg y_i : i \geq j - 1\},$$

and

$$y_j := y'_j \wedge \neg y'_j.$$

(3) In a further step, we see that the y_i (and thus the y_i') need not be different from each other, one y and one y' suffice.

Thus, we set $x_j := \bigwedge\{\neg x_i : i > j\} \wedge \neg y$, $y := y' \wedge \neg y'$.

(Intuitively, the cells are arranged in a circle, with y at the center, and y' "sticking out". We might call this a "curled saw blade".

(4) When we throw away the y_j altogether, we have Yablo's construction. this works, as we have the essential part in the x_i's, and used the $y_j's$ only as a sort of scaffolding.

Remark 7.6.5

It seems difficult to conceptually simplify even further, as Fact 7.6.3 (page 173) shows basically the need for the construction of the single Saw Blades. We have to do something about the teeth, and above Construction 7.6.2 (page 177), in particular cases (2) and (3) are simple solutions.

The construction is robust, as the following easy remarks show:

(1) Suppose we have "gaps" in the closure under transitivity, so, e.g. not all $x_0 \nrightarrow x_i$ exist, they always exist only for $i > n$. (And all other $x_k \nrightarrow x_l$ exist.) Then x_0 is still contradictory. Proof: Suppose x_0+, then we have the contradiction $x_0 \nrightarrow x_n \nrightarrow x_{n+1}$ and $x_0 \nrightarrow x_{n+1}$. Suppose x_0-, let $x_0 \nrightarrow x_i$. As x_i is unaffected, x_i+ is impossible.

(2) Not only x_0 has gaps, but other x_i, too. Let again x_n be an upper bound for the gaps. As above, we see that x_0+, but also all x_i+ are impossible. If $x_0 \nrightarrow x_i$, as x_i+ is impossible, x_0- is impossible.

(3) x_0 has unboundedly often gaps, the other x_i are not affected. Thus, for $i \neq 0$, x_i+ and x_i- are impossible. Thus, x_0- is impossible, as all x_i+ are, and x_0+ is, as all x_i- are.

Finally, instead of showing that two paths $\pi : x \ldots y$ and $\sigma : x \ldots y$ are contradictory, we may also show that all continuations $\rho : y \ldots z$, $\rho' : y \ldots z'$ have contradictions, $\tau : x \ldots z$ and $\tau' : x \ldots z'$ to $\sigma\rho$ and $\sigma\rho'$ respectively. Consider the following situation: $x_0 \nrightarrow x_1 \nrightarrow x_2 \nrightarrow x_3$, $x_0 \nrightarrow x_3$, $x_1 \nrightarrow x_3$, but $x_0 \nrightarrow x_2$ is missing, so we have no contradiction with $x_0 \nrightarrow x_1 \nrightarrow x_2$. $x_0 \nrightarrow x_1 \nrightarrow x_2 \nrightarrow x_3$ and $x_1 \nrightarrow x_3$ form no contradiction, as both legths are odd. We have a contradiction $x_0 \nrightarrow x_1 \nrightarrow x_3$ and $x_0 \nrightarrow x_3$, but, as x_1 is an "∨", we have to make sure that all continuations from x_1 have a contradiction with suitable $x_0 \nrightarrow x_i$, even without $x_0 \nrightarrow x_2$, they have to meet later on.

Diagram Simplified Saw Blade

Start of the saw blade before closing
the blade (without "decoration") under transitivity

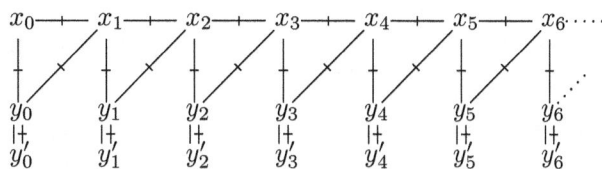

Diagram 7.6.3

Read $y_0 \Rightarrow_\pm y_0'$, $y_0 = y_0' \wedge \neg y_0'$, etc.

7.6.5   Paths Instead of Arrows

This section is not very systematic, and formulates some considerations for more complicated situations.

Recall Section 7.3.2.4 (page 146), discussing additional branchings.

Most of the basic reasoning here is analogous to the considerations in Example 7.3.4 (page 136).

7.6.5.1   A Generalization: Paths Instead of Arrows from x_0 to y_0

We now consider paths instead of arrows from x_0 to y_0.

(1) **Example 7.6.2**

See also Example 7.3.4 (page 136).

So we have contradictory paths σ, τ from x_0 to y_0. Suppose σ is the positive path (corresponding to $x_0 \not\rightarrow x_1 \not\rightarrow y_0$), and τ the negative one.

(1.1) Suppose σ does not branch. Suppose now x_0-, so we have $\vee$ at x_0, y_0 will be - and $\vee$, too, so stacking such contradictions does not help, we have an escape path downwards.

(1.2) Suppose σ branches, say at z, $\sigma : x_0 \ldots z \ldots y_0$.
Let x_0+

(1.2.1) if $z+$, too: We have the same situation as in Case (1) (constructed an escape route for the case x_0-), so this is not interesting.

(1.2.2) if $z-$: This situation is similar to the Yablo construction, and offers essentially nothing new.
The same considerations (infinity of branching, values $\pm$) as for the original construction apply.

(2) We generalize now Fact 7.6.3 (page 173) to paths.

See also Example 7.3.4 (page 136).

(2.1) The contradiction at x_0 cannot be $\sigma : x_0 \ldots y_0$, $\overline{\sigma} : x_0 \ldots y_0$ without branching by the argument against $x_0 \Rightarrow_\pm y_0$ in Example 7.3.4 (page 136).

(2.2) Assume by preprocessing that the first branching, say in σ, say in x_1, is equivalent to an $\vee$ for x_0+ (i.e. x_1 is an $\vee$, and σ up to x_1 is positive, or x_1 is an $\wedge$, and σ up to x_1 is negative) (otherwise contract the branchings to x_0).

(2.3) As we have $\vee$ at x_1, every path leaving x_1 needs a contradiction with some other path from x_0 (and the latter does not go through x_1, as a contradiction through x_1 would be invisible under the $\vee$ at x_1), no matter what the choices in other $\vee$ are.

(2.4) For x_0-, every path from x_0 has to lead to a contradiction, see again Fact 7.6.3 (page 173).

(2.5) Apply the iterated reasoning in Fact 7.6.3 (page 173).

7.6.5.2 Infinite Depth for Paths

We know that we need infinite depth, otherwise, we could fill in truth values from the bottom.

We give here a constructive argument.

See Section 7.6.3.2 (page 173) for details of the saw blade (and Yablo) construction. We go here into more detail.

Consider the construction $x_0 \not\to x_n \not\to x_{n+1}$, $x_0 \not\to x_{n+1}$. If x_0-, $x_0 \not\to x_{n+1}$ has to lead to a contradiction (recall, then $x_0\vee$), so we cannot stop at x_{n+1}, say we continue to x_m, then we have $x_0 \not\to x_m$, etc.

The case where the contradiction to $x_0 \not\to x_n \not\to x_{n+1}$ is not a simple arrow $x_0 \not\to x_{n+1}$, but a path $\sigma : x_0 \ldots x_{n+1}$ is more complicated, as there might be a contradiction already on σ. We now show that this does not work.

Say σ has the form $x_0 \not\to z \not\to w \not\to z' \not\to x_{n+1}$, with detour $z \not\to z'$, x_0+, $z-$, with an OR at z.

So we have a contradiction on σ, but not on the detour. But $x_0 \not\to x_n \not\to x_{n+1}$ has to be contradicted in all cases of the OR at z, so this will not work.

See Diagram 7.6.4 (page 182).

There is no easy way out. Suppose, by some construction between z' and x_{n+1} we could have a contradiction at x_{n+1}, i.e. $x_{n+1} -$. Then, irrespective of the value of z', we would have $x_{n+1} -$. But then we could append $x_{n+1} \to TRUE$ to x_{n+1}, and would have shown that z' cannot have a truth value by a finite construction. We know that this is impossible.

Of course, the same argument applies if we try to contradict $x_0 \not\to x_n \not\to x_{n+1}$ further down the road at some s, s' etc. $x_0 \not\to x_n \not\to x_{n+1} \not\to \ldots .s$ etc.

(By transitivity, we may take shortcuts, but never have only finitely many steps.)

Infinite paths

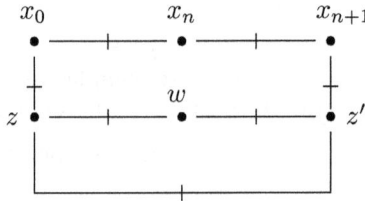

Diagram 7.6.4

Fact 7.6.6

We have infinite branching in Example 7.6.2 (page 180), too.

Proof

Recall the argument in Fact 7.6.3 (page 173).

(1) (Case x_0- in Fact 7.6.3 (page 173).)

   We need an ∇ at z' in $\sigma' : x \ldots z \to z'$ because of $x_0 -$.

(2) (Case x_i+ in Fact 7.6.3 (page 173).)

   (2.1) We may have a contradiction to $x \nrightarrow z \to z'$ by $x \to z'$.
   This works fine for $x+$, but not for $x-$, as we have then an arrow $x \to z'$, which does not lead to ∇ (as $z'-$), so we would have an escape path from $\vee$ to $\vee$.

   (2.2) As in Fact 7.6.3 (page 173), we have to introduce arrows $x \nrightarrow z''$ und $x \nrightarrow z'''$.
   We now have new arrows originating in x, we have to re-consider the case x_0- etc., as in Fact 7.6.3 (page 173).

□

7.6.6　Nested Contradictory Cells

Diagram 7.6.5 (page 184) illustrates combinations of contradictory cells.

We have a contradictory cell $\langle b, c, d \rangle$ in the left hand diagram, and may add new lines, forming additional contradictory cells, like the line $c-e$ in the central diagram, forming the cell $\langle c, d, e \rangle$, or the line $a-c$ in the right hand diagram, forming the cell $\langle a, b, c \rangle$, (These two possibilities are equivalent.)

There is a mutitude of possibilities, e.g. $a-d$, b-e, etc., we have not investigated, but we think they might not be very interesting - unless they form a nested construction like in or similar to Yablo's construction.

Diagram Nested Cells

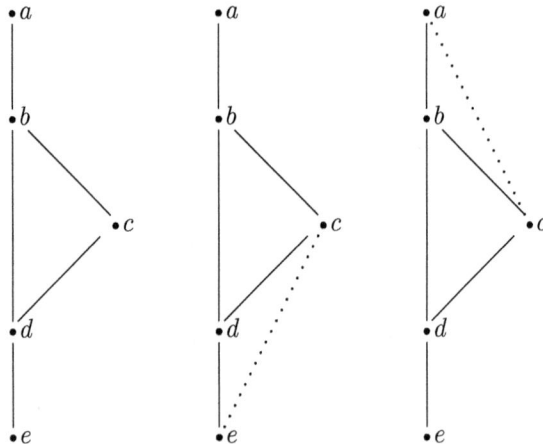

Diagram 7.6.5

Bibliography

[AA11]   M. Anderson, S. L. Anderson eds., "Machine Ethics", Cambridge Univ. Press, 2011

[AGM85] C. Alchourron, P. Gardenfors, D. Makinson, "On the logic of theory change: partial meet contraction and revision functions", Journal of Symbolic Logic, Vol. 50, pp. 510–530, 1985

[AIZ16]   A. Azulay, E. Itskovits, A. Zaslaver, "The *C*. elegans connectome consists of homogenous circuits with defined functional roles", PLoS Comput Biol 12(9), 2016

[Auf17]   "Aufmerksamkeit",   www.spektrum.de/lexikon/neurowissenschaft/ aufmerksamkeit/1072, 2017

[Avr14]   A. Avron, "What is relevance logic?", Annals of Pure and Applied Logic, 165 (2014) 26-48

[BB11]   A. G. Burgess, J. P. Burgess, "Truth", Princeton University Press, Princeton, 2011

[BBHLL10] J. Ben$-Naim$, J-F. Bonnefon et al., "Computer-mediated trust in self-interested expert recommendations", AI and society 25 (4): 413-422, 2010

[BP12]   J. Ben-Naim, H. Prade, "Evaluating trustworthiness from past performances: interval-based approaches", Annals of Math. and AI, Vol. 64, 2-3, pp 247-268, 2012

[BS17]   T. Beringer, T. Schindler, "A Graph-Theoretical Analysis of the Semantic Paradoxes", The Bulletin of Symbolic Logic, Vol. 23, No. 4, Dec. 2017

[CCOM08] R. Cabeza, E. Ciaramelli, I. R. Olson, M. Moscovitch, "Parietal cortex and episodic memory: An attentional account", Nat. Rev. Neurosci. 2008 Aug; $9(8) : 613 - 625$

[Chu07]   P. M. Churchland, "Neurophilosophy at Work", Cambridge University Press, 2007

[Chu86]   P. S. Churchland, "Neurophilosophy", MIT Press, Cambrige, Mass., 1986

[Chu89] P. M. Churchland, "A Neurocomputational Perspective", MIT Press, 1989

[DR15]   J. P. Delgrande, B. Renne, "The logic of qualitative probability", IJCAI 2015, pp. 2904-2910

[Dev65] P. Devlin, "The Enforcement of Morals", Oxford, 1965

[Dun95] P. M. Dung, "On the acceptability of arguments and its fundamental role in nonmonotonic reasoning, logic programming and n-person games", Artificial Intelligence 77 (1995), pp. 321–357

[Dwo82] R. Dworkin, "'Natural' Law Revisited", University of Florida Law Review, vol. 34, no. 2, pp. 165-188, 1982

[Dwo86] R. Dworkin, "Law's Empire", Cambridge, USA, 1986

[Ede04] Gerald M. Edelman, "Wider than the sky", Yale University Press, New Haven 2004, (German edition "Das Licht des Geistes", Rowohlt, 2007)

[Ede89] Gerald M. Edelman, "The remembered present", Basic Books, New York, 1989

[GLP17] "Gehirn und Lernen - Plastizitaet", www.gehirnlernen.de/gehirn/plastizitaet, 2017

[GR17]   D. Gabbay, G. Rozenberg, et al., "Temporal aspects of many lives", Paper 588

[GS08f] D. Gabbay, K. Schlechta, "Logical tools for handling change in agent-based systems" Springer, Berlin, 2009, ISBN 978-3-642-04406-9.

[GS10]   D. Gabbay, K. Schlechta, "Conditionals and modularity in general logics", Springer, Heidelberg, August 2011, ISBN 978-3-642-19067-4,

[GS16]   D. Gabbay, K. Schlechta, "A New Perspective on Nonmonotonic Logics", Springer, Heidelberg, Nov. 2016, ISBN 978-3-319-46815-0,

[Geg11] K. R. Gegenfurtner, "Gehirn und Wahrnehmung", Fischer, Frankfurt 2011

[HM17]   D. Hassabis, E. A. Maguire, "Deconstructing episodic memory with construction", Trends in Cognitive Sciences Vol.11 No.7

[Haa14] S. Haack, "Evidence matters", Cambridge University Press, 2014

[Hab01] J. Habermas, "On the pragmatics of social interaction", MIT Press, 2001

[Hab03] J. Habermas, "Truth and justification", MIT Press, 2003

[Hab73] J. Habermas, "Wahrheitstheorien", in Fahrenbach (ed.), "Wirklichkeit und Reflexion", Pfuellingen, 1973

[Hab90] J. Habermas, "Moral consciousness and communicative action", MIT Press, 1990

[Hab96] J. Habermas, "Between facts and norms: contributions to a discourse theory of law and democracy", MIT Press, 1996

[Han69] B. Hansson, "An analysis of some deontic logics", Nous 3, 373–398. Reprinted in R. Hilpinen, ed. "Deontic Logic: Introductory and Systematic Readings", Reidel, pp. 121–147, Dordrecht 1971

[Heb49] D. Hebb, "The organization of behavior", New York, Wiley, 1949

[Hem35] C. G. Hempel, "On the logical positivists' theory of truth", Analysis, 2:49-59, 1935

[IEP16] "Philosophy of Law", Internet Encyclopedia of Philosophy, 2016

[KLM90] S. Kraus, D. Lehmann, M. Magidor, "Nonmonotonic reasoning, preferential models and cumulative logics", Artificial Intelligence, 44 (1–2), pp. 167–207, July 1990.

[KPP07] B. Konikov, G. Petkov, N. Petrova, "Context-sensitivy of human memory: Episode connectivity and its influence on memory reconstruction", Context 2007: 317-329

[Key21] J. M. Keynes, "A treatise on probability", London, 1921

[Kri75] S. Kripke, "Outline of a Theory of Truth", The Journal of Philosophy, Vol. 72, No. 19, 1975, pp. 690-716

[LMS01] D. Lehmann, M. Magidor, K. Schlechta, "Distance semantics for belief revision", Journal of Symbolic Logic, Vol. 66, No. 1, pp. 295–317, March 2001

[Leh96] D. Lehmann, "Generalized qualitative probability: Savage revisited", Proceedings $UAI'96$, pp. 381-388, Portland, Or, Aug. 1, 1996

[Lew73] D. Lewis, "Counterfactuals", Blackwell, Oxford, 1973

[MP13] S. Modgil, H. Prakken, "A general account of argumentation with preferences", Artificial Intelligence 195 (2013) 361-397

[Mak19] D. Makinson, "Relevance via decomposition: a project, some results, an open question", Australasian Journal of Logic 14:3 2017, see also "Sets, Logic and Maths for Computing" (third edition), Springer 2020

[Mil06] J. S. Mill, "On Liberty", New York, 1906

[Neu83] O. Neurath, "Philosophical papers 1913-46", R. S. Cohen and M. Neurath (eds.), Dordrecht and Boston, D. Reidel, 1983

[OL09] M. Okun, I. Lampl, "Balance of excitation and inhibition", Scholarpedia, 4(8) : 7467, 2009

[Pul13]   F. Pulvermueller, "How neurons make meaning: Brain mechanisms for embodied and abstract-symbolic semantics", Trends in Cognitive Sciences, 17 (9), 458-470, 2013

[RRM13] L. Rabern, B. Rabern, M. Macauley, "Dangerous reference graphs and semantic paradoxes", in: J. Philos. Logic (2013) 42:727-765

[Rau21]   J. Rauch, "The Constitution of Knowledge. A Defense of Truth", Brookings Press, Washington, 2021

[Rot96]   G. Roth, "Das Gehirn und seine Wirklichkeit", Suhrkamp STW 1275, Frankfurt 1996

[Rus07]   B. Russell, "On the nature of truth", Proceedings of the Aristotelian Society, 7:228-49, 1907

[SEP13]   "Analogy and analogical reasoning", Fall 13 edition, Stanford Encyclopedia of Philosophy, 2013

[SEP19c] "Analogy and analogical reasoning", Stanford Encyclopedia of Philosophy, 2019

[SS05]   J. Sabater, C. Sierra, "Review on computational trust and reputation models", Artificial Intelligence Review, 2005

[Sab14]   K. J. Sabo, "Anankastic conditionals: If you want to go to Harlem ...", Draft for Semantics Companion, 2014

[Sch04]   K. Schlechta, "Coherent systems", Elsevier, Amsterdam, 2004.

[Sch18]   K. Schlechta, "Formal Methods for Nonmonotonic and Related Logics", Vol. 1: "Preference and Size", Vol. 2: "Theory Revision, Inheritance, and Various Abstract Properties" Springer, 2018

[Sch18a] K. Schlechta, "Formal Methods for Nonmonotonic and Related Logics", Vol. 1: "Preference and Size" Springer, 2018

[Sch18b] K. Schlechta, "Formal Methods for Nonmonotonic and Related Logics", Vol. 2: "Theory Revision, Inheritance, and Various Abstract Properties" Springer, 2018

[Sch18e] K. Schlechta, "Operations on partial orders", arXiv 1809.10620

[Sch95-3] K. Schlechta, "Preferential choice representation theorems for branching time structures" Journal of Logic and Computation, Oxford, Vol.5, pp. 783–800, 1995

[Sch97-2] K. Schlechta, "Nonmonotonic logics - basic concepts, results, and techniques" Springer Lecture Notes series, LNAI 1187, Jan. 1997.

[Sha13]   R. Shafer-Landau ed., "Ethical Theory", J. Wiley and Sons, 2013

[Sim63]   G. G. Simpson, "Historical science", in C. C. Albritton, "Fabric of geology", Stanford, 1963, pp.24-48

[Sta17a] Stanford Encyclopedia of Philosophy, "The coherence theory of truth", https://plato.stanford.edu/archives/fall2018/entries/truth-coherence (accessed 2017)

[Sta17b] Stanford Encyclopedia of Philosophy, "The correspondence theory of truth" https://plato.stanford.edu/archives/win2020/   entries/truth-correspondence (accessed 2017)

[Sta17c] Stanford Encyclopedia of Philosophy, "Epistemology" https://plato.stanford.edu/archives/fall2020/entries/epistemology (accessed 2017)

[Sta17d] Stanford Encyclopedia of Philosophy, "The Philosophy of Neuroscience" https://plato.stanford.edu/archives/fall2019/entries/neuroscience (accessed 2017)

[Sta18a] Stanford Encyclopedia of Philosophy, "Metaethics" https://plato.stanford.edu/archives/sum2014/entries/metaethics (accessed 2018)

[Sta18b] Stanford Encyclopedia of Philosophy, "Juergen Habermas" https://plato.stanford.edu/archives/fall2017/entries/habermas (accessed 2018)

[Sta18c] Stanford Encyclopedia of Philosophy, "The legal concept of evidence" https://plato.stanford.edu/archives/win2021/entries/evidence-legal (accessed 2018)

[Sta68]   R. Stalnaker, "A theory of conditionals", N. Rescher (ed.), "Studies in logical theory", Blackwell, Oxford, pp. 98–112

[Tha07]   P. Thagard, "Coherence, truth and the development of scientific knowledge", Philosophy of Science, 74:26-47, 2007

[WSFR02] P. Winkielman, N. Schwarz, T. A. Fazendeiro, R. Reber, "The hedonic marking of processing fluency: implications for evaluative judgement", in: J. Musch, K. C. Klauer eds., "The psychology of evaluation: affective processes in cognition and emotion", 2002, Lawrence Erlbaum, Mahwah, NJ

[Wik16a] Wikipedia, "Rechtsphilosophie",
            https://de.wikipedia.org/wiki/Rechtsphilosophie
            (accessed 2016)

[Wik17a] Wikipedia, "Memory",
            https://en.wikipedia.org/wiki/Memory
            (accessed 2017)

[Wik17b] Wikipedia, "Semantic memory",
            https://en.wikipedia.org/wiki/Semantic-memory
            (accessed 2017)

[Wik17c] Wikipedia, "Episodic memory",
            https://en.wikipedia.org/wiki/Episodic-memory
            (accessed 2017)

[Wik17d] Wikipedia, "Recognition memory",
            https://en.wikipedia.org/wiki/Recognition-memory
            (accessed 2017)

[Wik17e] Wikipedia, "Visual cortex",
            https://en.wikipedia.org/wiki/Visual-cortex
            (accessed 2017)

[Wik18a] Wikipedia, "Empathy",
            https://en.wikipedia.org/wiki/Empathy
            (accessed 2018)

[Wik18b] Wikipedia, "Diskursethik",
            https://de.wikipedia.org/wiki/Diskursethik
            (accessed 2018)

[Wik18c] Wikipedia, "Philosophy of science",
            https://en.wikipedia.org/wiki/Philosophy-of-science
            (accessed 2018)

[Yab82]   S. Yablo, "Grounding, dependence, and paradox", Journal Philosophical
            Logic, Vol. 11, No. 1, pp. 117-137, 1982

[ZMM15]   P. Zeidman, S. L. Mullally, E. A. Maguire, "Constructing, perceiving,
            and maintainig scenes: Hippocampal activity and connectivity", Cerebral
            Cortex, Oct. 2015, 25:3836-3855